MW01593139

PRUNING FOR
FLOWERS AND FRUIT

Dedication

To my axillary buds – sometimes opposite and sometimes alternate. I could not have written this without the constant support of my family, JV, Rosie and Jack Varkulevicius.

PRUNING FOR
FLOWERS AND FRUIT

JANE VARKULEVICIUS

CSIRO
PUBLISHING

CSIRO PUBLISHING GARDENING GUIDES

© E. Jane Varkulevicius 2010

All photographs and drawings copyright of the author unless otherwise attributed.

All rights reserved. Except under the conditions described in the *Australian Copyright Act 1968* and subsequent amendments, no part of this publication may be reproduced, stored in a retrieval system or transmitted in any form or by any means, electronic, mechanical, photocopying, recording, duplicating or otherwise, without the prior permission of the copyright owner. Contact CSIRO PUBLISHING for all permission requests.

National Library of Australia Cataloguing-in-Publication entry
    Varkulevicius, Jane.
    Pruning for flowers and fruit / Jane Varkulevicius.

9780643095762 (pbk.)

Includes index.
Bibliography.

Pruning.
Fruit – Pruning – Handbooks, manuals, etc.
Gardening – Handbooks, manuals, etc.
Flowers – Handbooks, manuals, etc.

631.542

Published by

CSIRO PUBLISHING
150 Oxford Street (PO Box 1139)
Collingwood VIC 3066
Australia

Telephone: +61 3 9662 7666
Local call: 1300 788 000 (Australia only)
Fax: +61 3 9662 7555
Email: publishing.sales@csiro.au
Web site: www.publish.csiro.au

Set in 10.5/14 Adobe ITC New Baskerville
Edited by Janet Walker
Cover and text design by James Kelly
Typeset by Desktop Concepts Pty Ltd, Melbourne
Printed in China by 1010 Printing International Ltd

CSIRO PUBLISHING publishes and distributes scientific, technical and health science books, magazines and journals from Australia to a worldwide audience and conducts these activities autonomously from the research activities of the Commonwealth Scientific and Industrial Research Organisation (CSIRO).

The views expressed in this publication are those of the author(s) and do not necessarily represent those of, and should not be attributed to, the publisher or CSIRO.

The paper this book is printed on is certified by the Forest Stewardship Council (FSC) © 1996 FSC A.C. The FSC promotes environmentally responsible, socially beneficial and economically viable management of the world's forests.

Mixed Sources
Product group from well-managed forests and other controlled sources
www.fsc.org Cert no. SGS-COC-003963
© 1996 Forest Stewardship Council

CONTENTS

ACKNOWLEDGEMENTS

With special thanks to Richard Aarons who kindly read through my drafts. His thoroughness, honesty and tact were most gratefully received and very much appreciated.

Many others were kind enough to help with their expertise and advice. Grateful thanks to all of you: Clifford Aarons; Richard Aarons; Greg and Jenny Bradshaw; Annmarie and Grayham Brookman, The Food Forest; David Button, Alameda Homestead Nursery; Nick Coulthard, Mount Mary Vineyard; Remy Favre and Blaise Vinot, Felco Distribution Pty Ltd; David Glenn, Lambley Nursery; Kath Kermode and Greg Daley, Daley's Fruit Tree Nursery; Bob Magnus, Woodbrige Fruit Trees; Frances Michaels, Green Harvest Organic Gardening Supplies; Becky Northey and Peter Cook, Pooktre; Phil Shepherd, Shepherd's Nurseries; Granville Parker, Mornington Peninsula Olive Oil.

INTRODUCTION

Why prune?

Pruning is all about how to control and direct plant growth. It is not brain surgery. Rarely are lives lost if a few simple rules are followed. Pruning is about how to bend plants to your will so you can make the most of every plant in your landscape, from fruit trees to groundcovers and grasses.

Anyone can prune; simply mowing the lawn is one form of pruning, but the results are more rewarding when you know what you are doing.

Armed with some pruning knowledge, when wielding secateurs or saw, you will make your plants more productive, more effective or simply more beautiful. Just like gentle discipline for children, good pruning should bring out the best in every plant.

Understanding how a plant grows greatly improves our confidence and effectiveness. When planting a new tree or shrub, how do you train it so it does not turn into a monster? When faced with a neglected tree or shrub, what is it the gardener sees? The tangle of branches? A plant too big for its space? These can be intimidating. 'Where do I start?' is the first despairing response, but the gardener is not 'seeing' the full story.

Plants are made up of two major components: roots and shoots. Roots anchor the plant in the soil by growing with gravity, that is, downwards. They act as food storage units (like carrots), but most importantly they draw nutrients and water from the soil.

Shoots (stems and branches) grow against gravity, upwards, carrying leaves with them into the light. The leaves interact with the light where, together with carbon dioxide and water, they manufacture simple sugars and oxygen.

As always, once you understand the system it is easy to manipulate it. Once the gardener has grasped how plants work, you can start to train them to suit your requirements.

Knowing when to prune to maximise growth or suppress it means that your site can hold more species than you originally thought, or that a screening hedge can be hastened into growth.

Sharp, well-cared for tools are essential for the finest finish on well-groomed plants, and will ensure that pruning for plant health is as effective as possible. By learning how to prune, many disease problems disappear, so toxic sprays can be dispensed with just by enhancing the amount of light and air available to the leaves.

Encouraging flowering growth and therefore fruit-bearing wood can maximise home harvests. Pruning for fruit requires the gardener to identify what growth their plants produce on, and how to keep the balance between the food-manufacturing leaves that will feed the hoped-for harvest.

If flowers are the priority, the same theory applies. Timing the pruning and encouraging flowering wood, rather than cutting it off, will naturally promote the most floriferous of gardens.

Virtually no gardener starts to cultivate on a clean slate. There are always a few tough survivors on the site. Pruning can reinvigorate and renovate plants that have been long neglected. It can turn a tangled mess into a bower of beauty or create a bountiful harvest. Knowing how to prune can make the most of whatever plants you may have inherited with your garden space.

However, it is essential that the pruner knows how plant systems work and develop. For successful pruning, the gardener needs to understand how plants grow.

1
HOW PLANTS GROW

Cambium – the uniting force

Scratch a woody twig and you will find a bright green layer called the cambium. The two plant parts, roots and shoots, growing in different directions, are united by a complex 'plumbing' network. This is known as the cambium layer. This layer links the microscopic root hairs gathering soil nutrients and water, with the shoots and leaves manufacturing food (see Figure 1.1).

These two elements combine and are spread through the plant from the veins in every leaf to the tip of every root.

The cambium layer consists of vascular bundles made up of two distinct types of 'plumbing' or vessels (see Figure 1.2).

The xylem and phloem form the core of the root. The xylem takes up the water and the phloem takes up minerals from the soil via

Figure 1.1 The xylem carries water and moves in one direction – straight up from the roots – and exits as water vapour through the leaves. The contents of the phloem move osmotically in both directions, carrying nutrients from the roots combined with simple sugars manufactured in the leaves. Together they penetrate and sustain all living parts of the plant.

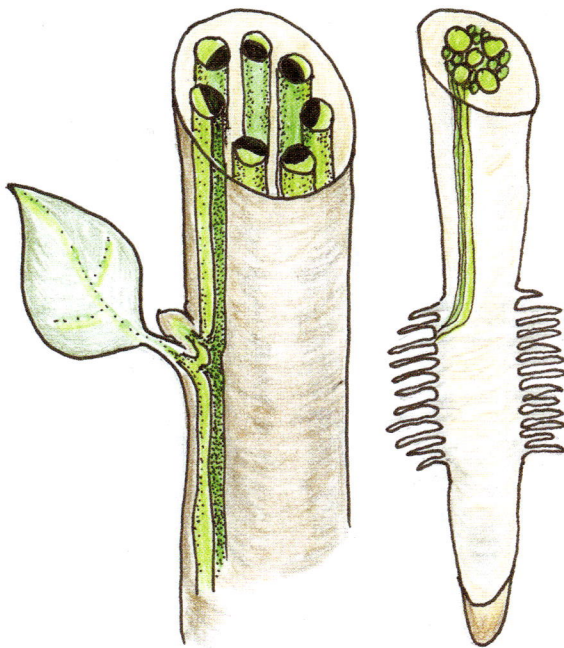

Figure 1.2 The cambium layer is made up of vascular bundles of xylem (dark green) and phloem (light green). The cambium layer, carrying nutrients up from the roots (right), and combining these with sugars manufactured by the leaves (left).

root hairs. The phloem also distributes sugars manufactured in the leaves to where it is needed in the plant. Together they penetrate and sustain all living parts of the plant. The xylem (pronounced *'zi-lem'*) is in charge of conducting water from the roots to the tip of the uppermost leaves – one-way traffic heading straight up and exiting through the leaves as water vapour. This water vapour is why it is often cooler in the shade of a broad-leaved deciduous tree on a warm day. The phloem (pronounced *'flow-em'*) carries sugars manufactured in the leaves to the whole of the plant, depending on where the plant needs nourishment. Well-nourished plants with well-nourished buds

produce more flowers and fruit than impoverished buds.

How the riches of the cambium layer are disbursed determines how well parts of the plant are nourished. This distribution of nourishment, and therefore growth, is determined by plant hormones that are active in growth points otherwise known as meristems (pronounced *'merry stems'*).

Hormones and meristems (points of growth)

As we all know, hormones are powerful things. Anyone that has lived with teenagers knows as much. They govern both the growth and development in all living things, including plants. A group of hormones called *auxins* govern which buds get nourished and produce growth, and which don't. Points of growth like buds are sites of active cell division stimulated with auxins (plant hormones) and are called *meristems.* Meristems will develop into buds producing leaves and wood, or flowers and fruit. Every seed/seedling starts with two meristems – the *radicle* and the *plumule* that give rise to all other growth points (see Figure 1.3).

As the plant grows, branching occurs. These branches/stems emerge from the growth points, meristems that develop after the germination stage. Their growth is governed by the concentrations of the plant hormones auxin and cytokinin.

These hormones are manufactured in the meristems (growth points) where plant cells are rapidly dividing to produce growth. They are found in the root tips and in buds.

Figure 1.3 This seven-day-old snow pea started with two meristems. The radicle forms the roots, and the plumule that forms the above-ground parts. Note the roots are branching and the seedling mix is clinging to the microscopic root hairs that draw up water and nutrients. The plumule is just unfolding its seed leaves that hide a meristem that gives rise to the rest of the future pea plant.

Buds – apical and otherwise

Buds form the above-ground growth points on plants. They contain the actively dividing cells and plant hormones that produce growth. Due to the balance of auxins, however, not all buds were created equal.

Apical buds

These buds are those at the very top (apex) of the plant or plant stem and produce auxins that keep this top bud extending towards the maximum amount of light available. Auxins also inhibit the growth of side or axillary buds. The apical bud, fuelled with hormones, gets the priority grab of water, nutrients and sugars required for growth. This means that it can grow taller,

bask in the light and out-compete the rest by suppressing those buds below it, creating what is referred to as *apical dominance*. Wherever there is a stem with an active apical bud going straight up, it will continue to do so regardless of the buds below it. There are plenty of people out there who are like this, and needless to say, they (the apical buds) should be trained to make sure that the whole plant can realise its maximum plant potential (see Figure 1.4).

The apical bud is essential for creating a strong central leader (trunk) in trees or other woody plants; however, it needs to be discouraged to form bushy plants. Successful lavenders would never have a central trunk, as their talent lies in forming a rounded shape in order to carry more flowers. Lavenders are a shrub, so are naturally inclined to be bushy. Pinching out the apical bud will ensure that the bush will branch from the base. This is often done in the nursery as soon as the

Figure 1.4 The most active growth is at the apex of the plant governed by the apical bud that pushes onwards and upwards. This is essential for a central trunk in trees.

cutting has struck. Groundcovers are plants that naturally grow horizontally. However, the density of the branchlets and subsequent leaves will be increased if the apical buds (on the end of the main stems) are pinched out. This will make the plant much more effective as a groundcover.

Axillary buds

To create bushy plants the solution is easy; just pinch out the top, apical bud, and there will be an immediate reallocation of plant resources to promote axillary or lateral (side) buds (see Figure 1.5).

Axillary buds, active or inactive, are found on the sides of the stems at the leaf nodes. They originate from meristems located between the base of leaves and the stem. They are the poor and completely oppressed, until stimulated into growth by a change in hormone levels brought about by removing the apical (top) bud. These axillary buds will one day form the side branches of the plant.

Figure 1.5 When the apical bud is removed, the meristems that have been suppressed in the leaf axils now become active. This is essential to encourage a bushy plant.

WHY PRUNE?

- Pruning can direct how water and nutrients are distributed through the plant
- It maximises the potential of certain growth points/meristems
- By understanding how the cambium and hormones work, the pruner can direct growth to where it is needed
- The apical bud will always dominate the rest of the plant, creating vegetative growth heading upwards
- Axillary buds growing sideways produce more flowering/fruiting wood

It stands to reason that the more light and nutrients a bud/node/leaf receives, the more active and productive it will be. The resulting healthier, stronger buds are more likely to produce flowers and/or fruit rather than the vegetative growth produced by the apical bud.

As branches grow more horizontally or sideways (lateral growth), the flow of the cambium (xylem and phloem) slows, rather than racing to the top of the plant. As the flow of the cambium slows, each bud/node has a greater opportunity to reap its riches. Thus the plant will produce more flowers and fruit on lateral growth than on vertical vegetative growth.

Not only are the hormone levels changed when the apical bud is removed, the axillary buds/nodes now have more access to light. This means they are more likely to develop leaves and stems which then photosynthesise, thus increasing the plant's food-manufacturing opportunities.

How plants make their own food

Photosynthesis is the process where the green parts of the leaf (chloroplasts) interact with light energy from the sun, carbon dioxide (CO_2) from the atmosphere and water via the xylem to create simple sugars. There is no life on Earth that does not depend on this process.

The chemical equation is:

$$6CO_2 + 6H_2O + \text{light energy} = C_6H_{12}O_6 + 6O_2$$

which means:

Six carbon dioxide molecules plus six water molecules plus light energy creates one molecule of sugar and six molecules of oxygen.

This simple sugar, combined with nutrients and water from the soil or growing media, supplies all the materials to make the plant and all other life on Earth.

The exposure of leaves to light is imperative. Look at the inside of a dense shrub; there are no leaves (see Figure 1.6).

Note the small bumps on the stem of the box plant in Figure 1.6. These are old meristematic sites (from leaves or small stems) that have gone into dormancy due to lack of light. Cutting back to these growth points and allowing light in can reactivate the meristems creating new growth.

Only the external parts of the shrub that receive light produce leaves (see Figure 1.7).

Photosynthetic surfaces may not always be what we think they are. It is the green pigments found in chloroplasts that perform photosynthesis. Chloroplasts are found not

Figure 1.6 The inside of this dense *Buxus* gets hardly any light, therefore it cannot support leaves.

Figure 1.7 The outside of the bush is well clothed with leaves due to their exposure to the light.

only in leaves, but in green stems and unripe fruit. These plant parts must be exposed to light to photosynthesise and therefore manufacture simple sugars.

Leaves and other plant parts

What, you may ask, does a grey or even black leaf do? How do cacti work, or 'air' plants with no roots at all? One thing is certain, tear any leaf or stem and you will find some 'green' inside. These are the green pigments in the chloroplasts that manufacture food by photosynthesis. Cacti have made their stems into photosynthetic surfaces, while the 'leaves' are the spines. Air plants, such as Spanish moss that hangs from trees in the Deep South of the USA, get their water from the humid atmosphere and their grey furry leaf coating limits moisture loss. The variety of plant physiology and adaptation to climate is mind-blowing; however, all plants have in common the ability to photosynthesise. The only exceptions are plants that draw nutrients and water from another plant by attaching themselves to the other plant's roots or stems – known as parasites.

The delivery of water, nutrients and sugars through the plant follow prescribed paths. Fortunately, the leaves and the buds they spring from are produced in just such an orderly and efficient way.

Leaf arrangement

The arrangement of leaves/buds on a stem are genetically predetermined in each species of plant. It is called *phyllotaxis*, from the Greek *phyllon* meaning 'leaf', and *taxis* meaning 'arrangement'. On many plants, such as

Figure 1.8 View from the end of an apple stem. Each leaf arranges itself so as not to shade the next.

apples and pears, the leaves are arranged in groups of five spiralling around the stem (see Figure 1.8).

When looked at 'end on' it is easy to see that each leaf is exposed to the maximum amount of sunlight while still not shading its neighbour. There are five buds perfectly angled until the next bud is placed directly above or below the first bud, but at some distance. Other plants have buds/leaves that sit opposite each other on the stem (see Figure 1.9).

While this may all seem interesting but not particularly relevant to pruning, consider that each leaf shelters a bud from which a stem may result. The leaf arrangement, especially in many large woody plants, dictates the placement of branches. It empowers the pruner to choose which direction the future branch will grow. This is

Figure 1.9 When leaves are arranged opposite to each other on a stem each pair angles away from the previous pair. This was taken in the afternoon and so the westward leaves get the maximum sun exposure.

LEAVES AND OTHER GREEN PARTS OF PLANTS

- They manufacture food for the plant and, ultimately, all of us
- They need light together with water and CO_2 to produce sugars, plant building blocks
- The base of leaves shelter buds/growth points/meristems that can become stems or branches
- Leaves are arranged in a predictable order (according to species)
- A pruner can encourage growth in a particular direction by cutting to a bud or leaf node facing in the desired direction

Your site and plant selection

The success of any landscape is reliant on the suitability of the plant material selected for the site. Always analyse the site before you plant to ensure quick establishment and bountiful fruit flowers and foliage. This may seem a long way from pruning but well-selected plants will reward the pruner. Those growing outside of their ecological range will struggle. There is nothing more depressing than dead plants, and they certainly don't need pruning!

There are a few basic site evaluations that need to be made before you plant. Never underestimate the value of local knowledge. Ask around and seek out local climate information.

Aspect

The site's aspect or relationship with the equator is crucial. It governs how much

imperative for the most elegant of espaliers (see page 131) for renovating old plants by cutting into old wood, directing the width of a canopy on weeping plants or while training any young plant. Note the leaf scars/dormant buds on the stems on Figure 1.6. Each leaf hides a bud/growth point/meristem ready to be stimulated into growth by pruning. See pages 54 and 83 for a list of which plants grow from old wood.

No matter how the leaves are placed on the stem, they are arranged so that each leaf will get its turn in the sun (see Figure 1.8).

sunlight will be available for your plants and all plants need light (see page 5). A southern hemisphere site that faces the north will get maximum sun, as will a south-facing one in the northern hemisphere. Landscape sites that face away from the equator, south in the southern hemisphere and north in the northern hemisphere, will be shadier. An easterly aspect exposes plants to the gentle morning sun, while those plants facing the west will get the full blast of the day's hottest rays. Adjacent buildings and plantings will also influence the levels of sun and shade.

Consider your plants' requirements for sun, heat and shade and plant them accordingly.

Elevation

Elevation, or the height that the site is above sea level, will also influence the climate. The higher above sea level, the cooler the site will be with more likelihood of frost compared to the surrounding area.

Topography – wind and frost

The shape of the land will also influence its climate in relation to prevailing winds, wind tunnels and frost pockets.

Wind

Planting on top of a hill will expose plants to every wind that blows, while planting on the leeward slope away from prevailing winds will give some shelter. Strong winds are destructive forces that strip leaves of moisture, making plant establishment difficult. Your site might need the planting of windbreaks before any general planting can commence.

In urban areas, the built environment may channel winds away from or towards your site. They need to be considered.

Frost

Just as hot air rises, cold air sinks. The lowest parts of a site will collect cold air that can turn to frost. The base of a hill will be much colder and frostier than a point halfway up the slope as the cold air flows like a liquid to the lowest point. These frost pockets should be planted with species that cope well with frost or cooler conditions – it is all a matter of plant selection. In warm areas where there may not be enough chilling hours for some fruit trees (e.g. apples), this low point in the topography can be utilised.

Masonry walls and paving can store the day's heat, especially if these face the equator or west. This heat will be released at night, warming the surrounding air and reducing the chance or severity of frost. Swimming pools or large ponds will also store heat that will influence the night temperatures around them.

Soil

Soil is that amazing material that supports plant life. Plants take the majority of their nutrients from the soil solution. This solution is the water in the soil that carries dissolved minerals and nutrients that in natural systems are derived from the soil itself and any fertilisers or organic matter added to it. The quality of the soil to a large extent influences the quality of the soil solution.

In hydroponic systems, where plants are grown without soil, plants grow using the nutrients added to the water they are growing in, a highly artificial soil solution made up of industrially produced fertilisers. This system is highly efficient in terms of plant growth of a particular species as the soil solution can be tailored to the crop's needs. When we create a diverse landscape, however, be it a food or ornamental garden or a public open space, the soil will determine the plant selection and its relative success.

What is it?

Soil is made up of a few basic components that vary in proportion with every soil. Mineral particles of varying sizes that are the result of rock weathering make up the bulk of most soils. The predominant size of these particles determines the texture of the soil and also its nutrient-carrying capacity (see below). Organic matter both living and dead is a vital if all too small component of soil. It can influence soil structure (see below), water and nutrient-holding capacity as well as providing habitat for all sorts of micro-organisms. This organic life can range from earthworms to soil fungi to viruses. The remainder of the soil is taken up with water and air.

It is the balance of these four components that determines its suitability for plant growth. There is sure to be a suite of plants that prefer the particular soil you have. Soils are as varied as the plants they support.

Soil texture

Soil texture is determined by the predominant size of the mineral particles that make it up. The best way to assess your soil texture is to get a handful of it and feel it. Knead the soil in your hands and focus on how it feels. Sandy soils have large particles that can be easily felt individually; heavy clay soils have extremely small particles so that it feels smooth and sometimes silky, like plasticine. Most soils will be somewhere in the middle. If you can fashion the soil into a roll or ribbon that holds together easily, you have clay soil. If the soil holds together briefly and you can feel both gritty and smooth elements, it will be loam. If it does not hold a shape it will be sand (see Figure 1.10).

Clay soil has a better water- and nutrient-holding capacity compared to sandy soil; however, it may have a poor drainage capacity causing lack of air in the soil profile that can lead to water logging. Sandy soil, on the other hand, is perfectly drained (and more prone to drought) as water flows through it easily. Unfortunately, nutrients also find it difficult to remain in sandy soil as they wash through the gaps between the large particles.

These problems can be ameliorated with the plentiful addition of well-rotted organic matter, the addition of bentonite and zeolites to sandy soils and gypsum to clay soils.

Soil structure

Good soil structure is vital. It differs from soil texture in that it is how the soil particles stick together (or not) rather than the actual size of the particles.

Well-structured soil should form aggregates or crumbs of soil that are a mixture of mineral and organic particles clumping together (see Figure 1.10b).

Figure 1.10 Different soils have different structure and soil texture. A heavy clay (a) can be improved with gypsum and organic matter resulting in a well-structured soil with stable soil aggregates that hold water and nutrients and allow the easy penetration of roots (b). Gravelly sand cakes after rain and holds neither water nor nutrients well (c).

These particles bunch together, creating different sized spaces where water and air can be stored, and they also provide a site for nutrients to cling to. These are essential ingredients for the successful growth of plants. The size and stability of these clumps/aggregates determines how well water, air, soil animals and plant roots move through the soil. Where there are stable soil aggregates, water will be absorbed easily, air can percolate to the roots and plants can spend more of their energy on shoot growth rather than forcing their roots through hard soil. Aggregates should be stable after rain and *not* form a hard crusty top that prohibits penetration by anything but a sharp spade or mattock!

The strong 'glue' that holds soil particles together to form stable aggregates is decomposed organic matter or humus. So whatever soil you have, the addition of composts and manures can only improve it. These can be dug in, or on really concrete-like clays, laid under a mulch covering in autumn and dug through the soil profile by earthworms. Too much digging will destroy soil aggregates, so the use of rotary hoes will obliterate a soil structure, creating a soil that sets like concrete when it is dry and turns into an amorphous slurry when it is wet. It sounds contrary, but a bit of digging in of organic matter will reduce the need to dig in the future – but don't dig too much. Keep adding organic matter and earthworms will happily dig it through for you.

Soil pH

Soil pH sounds technical but it isn't really. It is a scale indicating the acidity or alkalinity of the soil. Caustic soda (pH 14) is alkaline and battery acid (pH 1) is of course acid – plants will grow in neither as they are at the extreme ends of the scale. Somewhere in the middle, from pH 5.5 to pH 7.5, will be satisfactory for most plants, although some plants prefer one end of the range to another. Ericaceous plants like blueberries, rhododendrons and azaleas prefer a pH at or below pH6 where iron is freely available. Brassicas (cabbage, broccoli) and beans are best at pH 7.5 to pH 8 where there is plenty of calcium. Most plants, however, can grow in a wide range of pH levels.

The pH of the soil influences the availability of nutrients to the plant (see Figure 1.11). Plants grow only as fast as the least available nutrient; that is, if all nutrients are available to the plant except, for example, iron as it is an alkaline soil, the lack of this one element will retard the growth of the plant in general. The range of pH 6 to pH 7.5 ensures maximum nutrient availability.

Figure 1.12 pH testing kits are freely available and easy to use. The soil on the left is alkaline (pH 9). The soil on the right is acid to neutral (pH 6).

A pH test is easy to perform (see Figure 1.12) and can quickly solve the mystery of why a certain plant is not thriving in the existing

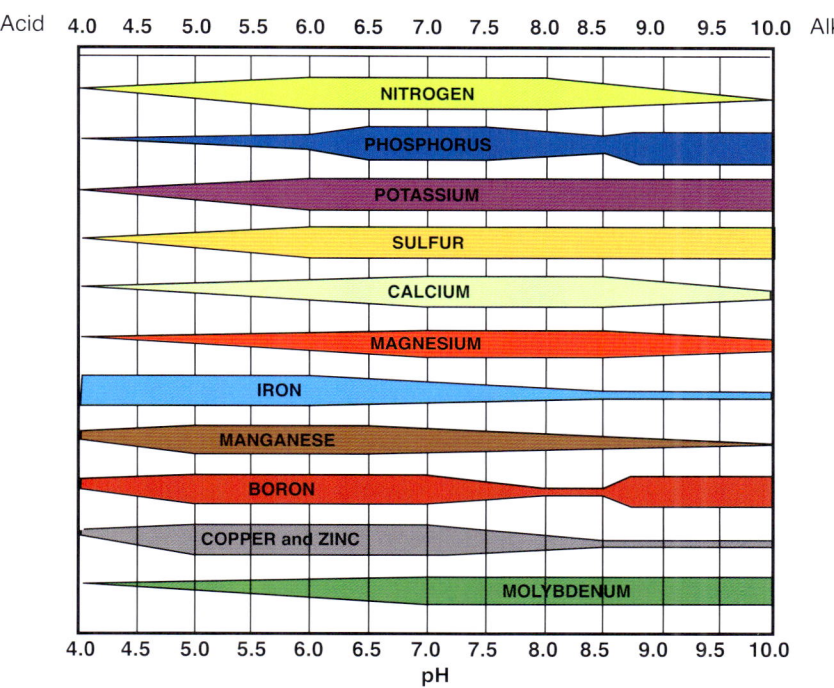

Figure 1.11 Soil pH determines how available nutrients are to the plant. The wider the band the more available the nutrient. (After K Handreck and N Black *Growing media for ornamental plants and turf*)

soil conditions. Overly acid soils can be corrected with the addition of ground limestone or dolomite, while alkaline soils can be treated with iron sulphate or sulphur.

Soil contamination

If all seems well with your soil, with the texture, structure and pH suitable for your selected plants and they still do not thrive, the soil may have been contaminated. Old industrial land may have a hidden residue of heavy metals or some other contaminant that will prevent plants growing. If you suspect this but don't want to have an expensive professional soil test, there is a quick and cheap way of finding out. Fill a seedling punnet with the suspect soil. Sow some radish seeds (they are very quick to grow) and see what happens. If the seedlings are deformed, badly discoloured or other than healthy, your soil may need professional testing. Soil removal is the only option for badly contaminated soil.

Buried rubbish in the form of old plaster or batteries will also affect plants as soon as they have made root contact. Dig down (carefully) and you may find the source of the problem.

Feed the soil to feed the plants

Depending on what you are growing, your plants may need some added nutrition in order to grow and fruit.

Fertilisers supply nutrients to the plants so that they have all the building blocks for health and growth. The major three nutrients are: nitrogen for vigorous leaf and shoot growth, phosphorous for photosynthesis and other chemical and hormonal functions, and potassium for stout cell walls to protect plants from disease. These three together with other mineral elements keep plants growing. There are many books written on fertilisers alone, and it is not the purpose of this book to examine them in detail. Suffice it to say, get to know your soil and the nutritional requirements of the plants you are growing.

Fertiliser use can be problematic with indigenous or recreated natural landscapes. It is rare on this crowded planet to find a soil that has not been altered by humankind. Agriculture has enriched soils for the preferred crops and introduced weed seeds abound in most soils. Many Australian natives abhor phosphates yet the farmland that is being gobbled up by housing developments was treated with superphosphate. This does not usually suit the indigenous plants as they have to compete with vigorous weeds and a possibly higher level of nutrients (or a particular one) than they require. Establishing natural grasslands also presents problems as they cannot compete with the introduced grasses whose seeds are stored in the soil. One solution is to strip the topsoil and plant into the nutrient poor and almost weed seed-free subsoil. If nutrient levels are too low, they can always be boosted.

Chemical, organic or mineral?

Whether your fertiliser is sourced from a factory or a farm, the nutrients it contains are taken up by the plants in exactly the same way. That is, they are taken up as chemical ions that are dissolved in the soil solution. NH_4^+ or NO_3^- are the same for the plant whether it is derived from cow manure or manufactured crystalline salts.

Chemically produced fertilisers dissolve rapidly into the soil solution and can 'burn' the plants with an overload of nutrients. The coated 'slow release' pellets will provide nutrients over a longer period but can be prone to 'dumping' their nutrients under certain soil conditions with the same result. Uncomposted manures can also be toxic to plants as the nutrient levels are so high they can also 'burn' the plants. Never use uncomposted manures and be sparing with powerful manufactured fertilisers. Adding that little bit extra 'just for luck' can have disastrous results, so always follow the manufacturers' instructions.

The major advantage of composts and well-rotted manures is that they also supply some organic matter that will improve soil structure (see page 9). Fertilisers derived from organic sources also have the full range of nutrients and trace elements in varying quantities, unlike the manufactured variety. This makes them simple to use in domestic and small-scale biodiverse contexts, though weed seeds may also abound. Be aware that some animals do not chew their food as well as others, and for this reason horse manure can harbour unwanted weed seeds. In commercial situations and monocultures where nutrient levels need to be carefully monitored, organic fertilisers may be too variable to be used with confidence. However, in most landscapes, using different sources of organic materials will deliver a satisfactory growth rate with the added bonus of improved soil structure that will boost water retention, drainage and the nutrient-holding capacity of any soil.

Mineral fertilisers consist of various crushed rocks and rock dusts that affect the pH and nutrient levels in the soil. Ground limestone and dolomite raise pH, as well as calcium and magnesium levels, gypsum helps break up heavy clays, and sulphur and iron acidify soils making other nutrients more available.

A successful landscape will only eventuate with a sound knowledge of the site and soil conditions and by selecting plants that thrive in that environment.

Photo taken at Southern Advanced Plants, Dromana, Victoria

2
PLANT QUALITY, PROPAGATION AND PERFORMANCE

The quality of your nursery purchase and how a plant is propagated influences how it will behave in your landscape.

Robust, well grown plants will establish easily, need practically no remedial pruning at planting, and if well cared for, will reward the gardener by reaching their maximum potential. Seek out reputable nurseries and pay that little bit extra to avoid possible disappointment in the future.

How a plant has been propagated or produced also influences a plant's performance and how it can be pruned.

A well-staked plant will have a strong root system anchoring it in the ground. Conversely, a badly staked plant can lead to a dangerously unstable tree that could be ever-reliant on its props.

The timing of your pruning will also determine the vigour of your plant, whether you need rapid growth or need to restrict a plant's size.

Choosing the right plant at the nursery

Do you pick the tallest because it looks the best value? How do you determine which plant will perform the best in your landscape? Most certainly it is not the tallest one!

Once you have decided which plants you want to grow, there are a few criteria to keep in mind when selecting a healthy plant that will establish well. This is most important when selecting trees or large woody shrubs. Smaller plants such as grasses and sub-shrubs are unlikely to be stressed and destabilised by wind, and in the case of herbaceous perennials, the roots and shoots are renewed annually. This is not so with trees.

Trees are long-term members of the landscape and, due to their size, they need to be structurally sound to resist high winds without losing limbs or falling. They also need vigorous root systems to take up water and nutrients from the soil to establish in their new homes and to anchor the plant in the ground.

The 'root to shoot' ratio

The most important consideration when selecting your plant is the health and size of the roots in relation to the size of the shoots (trunk, branches, leaves).

This is called the *root to shoot ratio*. In a perfect world the volume of the roots would be roughly equal to or exceed the volume of the aboveground parts. This is, after all, the pattern in well-established plants.

Most trees, however, are sold in containers and nurtured with ample amounts of water and nutrients so that the restricted roots can support a shoot system much larger than that which would occur when planted in the landscape. Trees with a greater volume of shoots in relation to the roots will find it hard to draw up sufficient water and nutrients to establish the plant quickly and successfully. The first thing to consider when choosing a plant is the size of the container (holding the roots) in relation to the top growth (see Figure 2.1).

'Bare-rooted' plants are sold when they are dormant and have no soil around their roots. The small, sturdy root system should be checked for any broken or damaged roots. Trim them back to reduce the likelihood of infection. Take care to preserve as many fibrous roots as possible (see Figure 2.2).

The shoots of these plants may need to be pruned back when they are planted to compensate any for discrepancy in the ratio between roots and shoots. Sometimes the shoot growth has been pruned back, but often to an inward-facing bud. This is so the outward-facing buds are less likely to be

Figure 2.1 These are very tall trees for the size of their pot. The root system is tiny compared to the amount of shoot growth. It is unrealistic to expect these trees to establish easily with such a small root system.

damaged in transport to the nursery. Make sure you prune the stems to an outward-facing bud so the centre of the plant will not become crowded (see Figure 2.3).

Figure 2.2 A bare-rooted tree from the nursery.

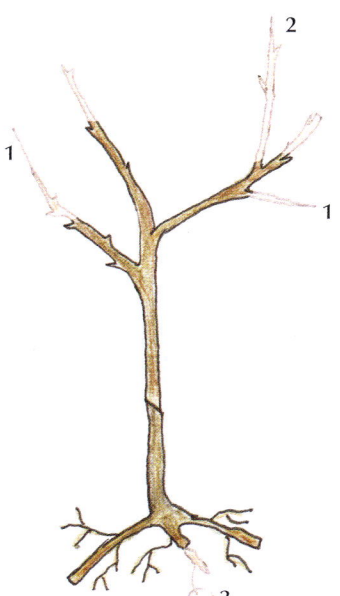

Figure 2.3 How to create an 'open vase' tree. Prune to an outward- and downward-facing bud and remove weak growth (1). Remove any upright growth that may compete with the trunk (2). Cut back damaged roots (3).

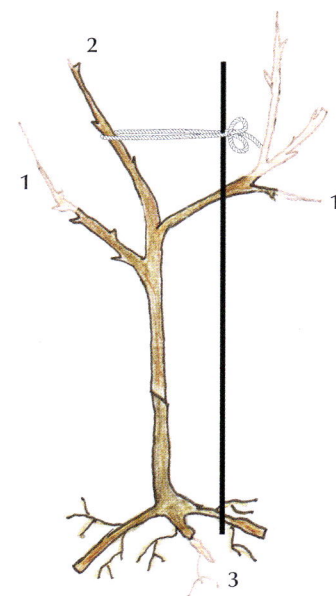

Figure 2.4 How to create a spindle or pyramid tree. If there is no plant with a central leader, train the uppermost branch to a vertical position. Tie the branch with soft ties and bend it into position. Fasten the ties to a stake that holds it in position. Prune to an outward- and downward-facing bud and remove weak growth (1). Aim to direct side growths to the horizontal where possible. Train the dominant upright branch to the vertical (2). Cut back damaged roots (3).

State of the roots

Never be afraid to inspect the root system where possible. It is, after all, an essential part of the plant purchase! In some cases there is evidence of root spiralling where the roots have circled round the inside of the container (Figure 2.5).

This situation must be rectified for the plant to succeed. It will mean a check in growth and slower establishment. If there is no alternative to this pot-bound plant, the bottom 2 cm of the root ball can be removed and the sides of it slashed at planting (see Figure 2.6).

Figure 2.5 Spiralling roots, if not corrected, will never be able to venture out into the surrounding soil. The plant is unlikely to establish well.

Figure 2.6 The spiralled roots have to be slashed and much of the bottom roots removed so that new roots can explore the surrounding soil. The plant will need some time to establish well.

This will allow the roots to extend into the soil it is planted in. If this step is missed, the roots will continue to circle, depriving the plant of water and nutrients in the soil. The plant will also be unstable.

Root girdling is almost impossible to detect. It occurs when the young seedling is potted up and the seed root (radicle) is bent up like a 'U' as it is pushed into the potting mix. As the plant grows, the U-shaped root grows to effectively restrict the flow of water and nutrients to the rest of the plant. At some time in the future the plant is likely to snap off at this point. It is a problem associated with fast-growing trees such as eucalyptus. Often the plant feels unstable in the pot when rocked. Otherwise, just steer clear of bargain basement plants and buy such long-term investments from the most reputable growers. *The extra time and expense taken can mean the difference between success and failure in the landscape.*

Root-training pots and air pruning

To overcome the problems outlined above, many nurserymen are using different containers. Many have ridges down the inside of the pot to inhibit spiralling (see Figure 2.7).

The roots hit the ridges in the pot and are guided downwards. When the roots reach the bottom of the pot and come into contact with the air, they become thicker and knobbly. This thickening is a store of carbohydrates that will fuel rapid root growth once planted.

In more advanced plants, the use of spring rings and root control bags allow air pruning of the roots, resulting in a mass of fibrous white roots. These vigorous roots will

Figure 2.7 Root-training pots.

Figure 2.8 Masses of fibrous white roots mean quick establishment in the landscape.

establish themselves quickly once transplanted (see Figure 2.8).

Branches

Branch placement also needs to be considered. Trees with branches growing at a wide angle from the trunk are strong. Those growing at very close angles or even parallel to the trunk will lead to long-term instability if not pruned out at planting (see Figure 2.9).

Branches should also be widely spaced around the trunk, rather than all originating from a small section (see Figure 2.10).

This poor branch configuration which was not corrected early in life has forced the owner to remove the tree. Some of the angles

Figure 2.9 Two competing leaders such as these are a recipe for disaster.

Figure 2.10 Too many branches originate from too small an area of trunk, even though some of the angles are wide and stable. Early pruning could have prevented this tragic result.

Figure 2.11 This advanced birch tree has been fastened to a wire in the nursery. The subsequent damage and swelling makes this a weak point on the tree.

might be right but the trunk cannot support so many branches in almost the same spot. The pressure exerted in high winds is too great to be sustained and eventually branches are lost, hopefully without harming people or buildings. Engineering in trees is just as important as it is in buildings.

Trunks

The thickness of the trunk (or trunk calliper) is also a good indicator of ultimate stability. Trees with thick trunks in relation to their height have generally been grown at wider spacings in the nursery, allowing for a denser canopy of leaves that will support speedier root growth and therefore quicker establishment. A short and stocky tree with a low centre of gravity may not be elegant, but it is stable. Just think of those rocking children's toys that refuse to be pushed over.

The trunk should also be thickest at the base and then taper to the top. Some advanced trees have been fastened to wires to keep them upright. Often the thickest part of the trunk can be at the point it was tied to its support. This can also be a point of weakness in the future (see Figure 2.11).

PLANT QUALITY

- Good selection is most important with trees
- Right proportion of roots to shoots
- Roots should not spiral
- Healthy roots are white and fibrous
- Plant is stable in its pot
- The trunk tapers evenly from bottom (widest) to top
- Branches evenly spaced
- Wide angles between the branch and the trunk

Propagation and landscape use

A plant's manner of propagation will determine its landscape use and how it should be pruned. Plants can be propagated sexually or asexually; that is, they can be grown from seed, a product of sexual union (pollination) between pollen (male cells) and ovules (female cells), or grown asexually by cuttings, grafting or budding, or from tissue culture.

Asexual propagation results in plants that are genetically identical, therefore they will perform in a uniform manner given the same growing conditions.

Seed grown plants – sexual propagation

As a result of sexual reproduction, seedlings are as variable as we are. This has advantages and disadvantages. In recreated natural landscapes, genetic diversity is of primary importance so only seed-grown plants should be used. If an avenue of trees or a hedge is being planned where uniformity of growth is required, plant asexually propagated plants; that is, clones.

Seedling-grown plants have a typical tap root system that anchors them in the ground (see Figure 1.3, page 3). In large woody plants these root systems are easily damaged with dire consequences, especially with eucalypt species. Care must be taken when selecting such plant material (see page 18) as the root-to-shoot ratio is critical in the successful establishment of seedling-grown large woody plants.

Pollination

Pollination that results in seed/fruit production can only occur between members of the same species, although there are exceptions. Most plants have perfect flowers; that is, they carry both male and female parts in the same flower. They may need cross-pollination from another of the same species or can be self-pollinating/self-fertile.

Some plant species are *monoecious*. This means that male flowers and female flowers are produced separately, but both on the one plant ('mono' meaning 'one'). Pumpkins, squash, sweet corn, hazels and walnuts are all examples of this (see Figure 2.12).

Figure 2.12 Corn is monoecious; the male flowers at the top drop pollen on the cornsilk stigmas that lead to the corn's ovaries below.

Other plants are *dioecious*. These species produce male flowers on one plant and female flowers on another. Therefore, it is necessary to have as least one female plant and one male in order to achieve fruit production. Carobs, hollies and pistachios are all dioecious. Only one male plant is required to pollinate about five female plants but this can depend on species and climate. Suffice it to say, if you are after fruit production, plant more females than males, or if you wish to avoid fruit/seed production, plant only males. This is often the case with *Ginkgo biloba*, the maidenhair tree whose fruit smells disgusting – not a desirable attribute in an ornamental tree, so male plants are preferred. Unless the plants are produced asexually, we are not technologically advanced enough to tell male from female at seedling stage. Adherents of geomancy and dowsing have their theories that may or may not be useful.

Seedlings that 'come true' from seed

Some plants, many of them vegetables, are open-pollinated such as tomatoes, lettuce and beans. They have closed flowers and therefore pollinate themselves. No other pollen apart from their own can enter the flower and cross-pollinate with it. These are our inheritance of millennia of selective breeding by farmers and gardeners.

Some seed-grown plants have been cultivated for so long that even though they are the same species, variations of the species have been selected for different purposes. The species *Brassica oleracea* is a prime example. Cabbages, cauliflowers, broccoli and Brussels sprouts are all the same species. These variations have been selected and interbred resulting in very different-looking plants (see Figure 2.13).

The cauliflower and the cabbage in Figure 2.13 can be grown from seed and 'come true from seed' as the saying goes. This is only because of millennia of seedling selection and how the seed was produced; for example, red cabbages were only allowed to be

Figure 2.13 Cauliflower (bottom) and red cabbage (top) are the same species, *Brassica oleracea*. Through selective breeding, they have evolved into very different-looking plants.

pollinated by other red cabbages. This is done by ensuring that no other *Brassica oleracea* pollen can dilute the red cabbage pollen. As bees have a range of at least 2 km, fine mesh is used to cover the flowering plants to prevent cross-pollination from (for example) a flowering Brussels sprout.

Hybrid and genetically modified seeds do not 'come true' when saved. They have to be sourced fresh each season from the multinationals – a goldmine for these companies, expensive and usually unnecessary for us.

Asexual propagation – cutting grown and grafted plants

Asexual propagation is all about producing clones of a particular plant, creating a population of genetically identical plant material. This is the story of the cultivar. A cultivar is propagated without change, therefore by asexual means.

The Granny Smith apple found in almost every fruiterer is the fruit of a clone; that is, the tree from which it was picked is genetically identical to the first chance seedling that grew and fruited in Mrs Thomas Smith's garden outside Sydney in 1860. This is achieved by grafting a piece of the original tree (scion) onto a rootstock (see Figure 2.14).

The Granny Smith apple in Figure 2.14 on the left is made up of a rootstock (**1**) and a piece of the original Granny Smith tree (**2**), the scion. The standard rose on the right is made up of three different parts. The rootstock that forms the roots and the trunk (**1**), with two different roses grafted onto to it; one pink (**2**) and one white (**3**).

Figure 2.14 Two examples of grafted plants.

The snowball tree, *Viburnum opulus sterile*, is, as its name suggests, sterile and produces no seed. It is grown from stem cuttings as it can grow adventitious roots. Adventitious roots do not arise from a seed but from a piece of stem, leaf or branch inserted into a suitable growing media. This has advantages. An adventitious root system, instead of growing a vertical tap root, produces many roots that grow more horizontally. This allows better access to nutrients in the richest top section of the soil profile and also indicates an ability to recover from root damage.

Implications for pruning

Grafted plants are in fact two to three plants on the one root system. If the plant is cut off below the graft or throws up suckers from the root system this growth *will not be* of the desired cultivar. It should be removed or regrafted with the desired scion wood.

Plants grown from cuttings are one plant. This means that if the plant is decapitated or throws up suckers from the root system, it will still be the desired cultivar that was purchased. It will also be genetically identical to other plants of the same cultivar so uniformity of growth for hedging or avenues is assured.

Seedling-grown plants are variable. They are essential for biodiversity as each seedling is an individual with differing strengths and weaknesses, just like human beings. Their flowers, bark, branch habit and disease resistance will vary in each individual. This is why there is strength in diversity. A genetically diverse plant base can overcome changes in the landscape or climate. Not all will survive but there will be individuals that have the genetic traits required for coping with differing conditions.

However, in ornamental landscapes where uniformity of growth is desirable, never use seedlings. The same can be said for food plants such as fruit trees. Fruit trees grown from seed are unreliable. They may take up to 20 years to fruit, in the case of avocados, and the fruit may not be worth gathering when it is produced. Take advantage of our agricultural heritage of millennia of selective breeding. Purchase a known cultivar for guaranteed fruit production.

Staking plants

Staking plants has become a bad habit in horticulture and rarely, if ever, is a well-staked plant seen. There are a few good reasons for staking plants, but this technique is over-used to the detriment of the plant material concerned. Like over-indulged Western children who are not taught to be self-reliant, a plant that is used to being propped up will become unstable when the supports are removed. Most plants do not need staking and develop stronger root systems and better height-to-girth ratios when unstaked. Just as the best tree to choose at the nursery is wider than it is tall (see 'Choosing the right plant at the nursery' on page 15), a tall skinny tree that develops as a result of over-long poor staking is unstable.

Advanced trees will need staking as their root system needs to be kept stable in the soil. A stable root system that is not rocked by winds

Figure 2.15 How to stake a tree properly.

will be able to send out roots into the surrounding soil to anchor the plant firmly for the future. *Securing the roots is the object of staking, not keeping the trunk rigid.* Trees on dwarfing rootstocks should also be staked initially as the dwarfing rootstock is slow growing and will need some time to establish. A three-point tie in the lower third of the trunk is advisable for the first year or 18 months, then the stake should be removed (see Figure 2.15).

Never stake as high up on the trunk as shown in Figure 2.16. The tree will be forever dependent, and weak once the stakes are removed. The trunk needs to be able to

move so that it can develop some thickness and strength.

Topiary standards (see page 101) will need to be staked in the training stages, and if a large 'ball' is developed making the plant top heavy, constant staking is advisable. If it is grown from the one plant, that is not a grafted standard, and the 'ball' is not overlarge, the stake can be safely removed at the end of training (see Figure 4.75, page 101). Grafted standards such as roses or grevilleas will, however, need to be staked for life. The grafting point at the top of the trunk is a weak point that can snap in high winds.

Figure 2.16 A badly staked tree.

Some nursery stock that has been grown fast and has not developed a self-supporting trunk will need a slender stake to keep the wobbly stem/future trunk straight. This is most notable in olive trees. Don't overfeed the tree and remove the stake as soon as the trunk is sufficiently thick as to remain upright without support.

An unfastened stake can be useful in re-created natural landscapes merely to mark the spot where the tree is planted. It can prevent mower damage when clearing long grass. However, most trees and shrubs, if well selected at the nursery, will need no staking.

When to prune

The timing of pruning influences the vigour of the new growth on your plant, whether it is vigorous or moderate. Prune a flowering shrub at the wrong time and you can remove all your flowers. It can make your plant more susceptible to frost or increase or decrease your chances of disease. Pruning times for various types of ornamental plantings are covered in 'Ornamental plants' on page 49. Pruning deciduous trees, including fruit trees, can be timed differently depending on the desired results.

Winter pruning

Generally most pruning of deciduous trees has been done in winter when the plants are dormant. In winter the framework of your plant is clear. It is easy to see diseased and damaged wood, or crossed branches that are impeding air flow through the canopy. These all need to be removed as they provide a haven for disease.

Pruning in winter will stimulate vegetative growth. Over the growing period, in good seasons, the tree stores carbohydrates in its trunk and roots to allow the plant to survive through the winter without leaves ('How plants make their own food', see page 5). The tree will have stored enough to keep the summer-sized tree alive. If there is heavy pruning in the winter (that is, it is no longer the same sized tree it was in autumn), there is an oversupply of stored food and plenty of strong growth will occur the following spring. This can be desirable to re-invigorate and produce new wood on old or weaker plants, but not if you wish to limit the tree's size. The growth produced after winter pruning tends to be vigorous and vegetative wood rather than wood that will carry flowers and fruit.

Pruning when your plant is dormant also means that pruning wounds will not heal as quickly as they would if it was in vigorous growth. As all deciduous trees are in effect in hibernation in winter, expending only enough energy for the tree to continue respiring, there is not the quick response healing that would be expected when it is actively growing. Pruning wounds are the obvious starting point for disease to enter your plant, and winter pruning creates wounds when fungal spores are more prevalent. Climate also plays a part in the timing of winter pruning. Young trees are more susceptible to damage in hard winters. Leave pruning to late winter, early spring or after the danger of frost has passed to minimise damage in frosty areas. When the tree is actively growing healing will be quick.

Summer pruning

In many landscapes, it is the restriction of overall plant size that is desirable and therefore summer pruning is most effective. By removing vigorous spring growth in summer, the pruner is also removing those vital food-producing organs – the leaves. With less food production, plant growth is restricted, and the tree's food reserves will come into balance at the height and width that is set in summer.

WHEN TO PRUNE?

- Winter pruning stimulates vegetative growth
- Pruning wounds don't heal as well in winter
- Summer pruning limits growth
- Summer pruning makes trees easier to net
- Summer pruning allows improved air circulation around the fruit
- Summer pruning can maximise fruiting wood

As the tree is actively growing, pruning wounds will heal faster, and on dry, clear days, fungal spores will find it more difficult to enter the plant.

It is also convenient for netting fruit trees. Pruning just before netting means there is not so much tree to cover, the net fits more snugly and air circulation is increased. This means that as cool autumn nights arrive with warm days, there is less likelihood of your fruit getting fungal infections.

In cooler areas, summer pruning also opens up the tree canopy to the sunlight so that fruit will ripen more evenly. In areas with hot summers, however, don't remove so much growth that it exposes the fruit to sunburn damage.

Once your tree is established, getting into the habit of summer pruning is generally best for both pruner and plant. Growth is more easily managed and venturing out on a cold winter's day need only be for removing any dead wood, or removing what has been missed in the summer.

3
TECHNIQUES AND TOOLS

The whole object of pruning is to make your plants as productive, attractive and healthy as possible, *so don't be afraid*. A few basic techniques are all you need to know for success.

Don't get overly fussed. Plants are geared to grow no matter what we do. The roadside fruit tree in Figure 3.1 has obviously never been pruned, weeded, watered or fed, yet it is covered in blossom.

It will bear masses of probably very small fruit, but the lack of pruning, or the attentions of local council pruning haven't killed it.

Figure 3.1 Productive, attractive, thriving – and never been pruned!

So what is the kindest cut?

The smaller the cut, the easier it is for the plant to heal. Pruning and training early in a plant's life means that there is less likelihood of having to make large corrective cuts later. Yes, again it is just like children – forming good habits when young creates a solid base from which to grow.

Wounds

All pruning inflicts wounds. However, the smaller the wound the easier they heal. To be exact, a wound does not 'heal'; it is overgrown by healthy tissue that has been stimulated into action by the wound. A clean cut leaves a smaller surface area for the plant to overgrow, so sharp pruning tools are essential (see Figure 3.2).

Conversely, the larger the cut, the longer it takes to heal (see Figure 3.3). Early pruning could have prevented this.

There is always much debate about whether large wounds should be covered in some sort of dressing. The natural desire to put a 'band aid' on a cut seems to extend from our human nurturing instincts; however, healthy plants are quite capable of healing themselves.

Figure 3.2 Small cuts heal easily and are quickly overgrown.

Figure 3.3 A large branch has been removed leaving a huge scar.

Knowing how to prune well is an advantage certainly, but dressings on plants are not only unnecessary but can be detrimental.

> Applying a dressing is likely to keep the wound moist, creating a perfect environment for the growth of disease carrying organisms. Keep plants healthy, prune wisely and wounds will heal/grow over by themselves.

Hygiene

Always disinfect your tools after pruning a particular plant and before pruning another. This will prevent disease spreading from one plant to the next. Dip or spray your tools with methylated spirits or household bleach and wipe dry with a clean cloth before use.

Rubbing off

Tools

The only equipment you need is your thumb.

Technique

Rub off unwanted growth when it is young or it will some day become a problem involving major surgery (see Figures 3.4 and 3.5).

If allowed to grow, unwanted buds will form a thicket of branches, crowding the canopy with unproductive wood.

Spring is the time to watch out for this sort of growth which is most prevalent after heavy winter pruning. Work out how these growths fit into the overall shape of your plant and edit them accordingly. Look at which way they are heading. Some of these growths will be facing in the direction you want, others growing inwards towards the centre of the bush, or too close together. These will crowd your plant and should be removed.

Always remove such young growth if it is occurring below the graft on a grafted plant. This will be growth from the rootstock that will eventually take over the desired clone that was grafted onto it. A rootstock is used to create a plant of a certain size, to increase

Figure 3.4 Unwanted buds will turn into unwanted, crowded branches. Deal with them early.

disease resistance or to cope with hostile soil conditions. It enables your chosen plant to function – it is not the chosen plant itself. Be ruthless (see Figure 3.6).

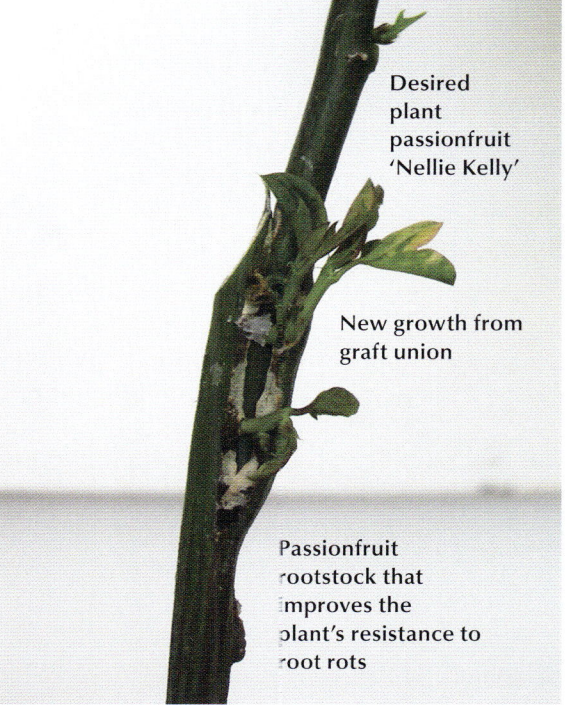

Desired plant passionfruit 'Nellie Kelly'

New growth from graft union

Passionfruit rootstock that improves the plant's resistance to root rots

Figure 3.5 A few rubs of the thumb could have prevented this superfluous growth.

Figure 3.6 New growth is sprouting from the actual rootstock. Rub it off now before it takes over the desired plant grafted above it.

Pinching out, tip pruning

Tools

A forefinger and thumb is all that
is required.

Technique

Pinching out the tip of a stem immediately
reallocates the plants internal resources
such as hormones and the flow of the
cambium layer (see page 1). Any plant or
part of a plant that you want to make
bushier benefits from this technique.
By pinching out the tip, or apical bud, the
buds below are stimulated so that you
promote two or three smaller stems rather
than one long stem, hence the plant is
bushier (see Figure 3.7).

Again, spring is the obvious time to look
over your plants and encourage bushy
growth at will. Pinching back new growth
on espalier trees is done in early summer
(see page 131) and some seedlings can be
pinched out as they grow to make a strong
framework. Annual and biennial flower and
vegetable seedlings also benefit so that they
develop more side branches from which to
carry flowers (see Tables 3.1 and 3.2).

Generally, leaf buds are thin and pointy
while flower buds are round and fat.
Pinching out flower buds in the early
stages of growth can be beneficial as
growth will be directed to the development
of roots and shoots. Once the plant has
become established, pinching out leaf
buds only will create more branches
from which flowers will be carried
(see Table 3.3).

Figure 3.7 Using your thumb and forefinger pinch out the top bud (a). The top bud has been pinched out leaving two side buds (b). The tiny side buds can develop and branch (c).

Table 3.1 Vegetables that benefit from pinching out

Amaranth
Broad beans – pinch out the top of the stem when the lowest flowers on the stem are showing signs of finishing. It will help 'set' the pods.
Broccoli sprouting, broccolini
Capsicum, chilli
Eggplant
Okra
Peas – when the vine reaches over the top of its trellis, pinch out the top shoot and use for stirfrys. This will also help set the pods.
Pumpkins, cucumber and melons – when the vine has grown to the edge of the garden space allotted or when there are four or five female flowers developing. Pinch out the ends of the vine; this will help 'set' the fruit.
Tomato

How to cut

How to cut – small stems up to 25 mm diameter

Tools

The best results are achieved with razor sharp secateurs. By-pass secateurs are the best. They work like a pair of scissors where the upper blade slides past the lower blade, which in turn holds the branch in place. This is the classic gardener's secateur.

Anvil secateurs are also available but work like a meat cleaver on a chopping block, and

Table 3.2 Herbs that benefit from pinching out

Basil
Coriander
Dill
Lavender
Mints including Vietnamese mint
Oregano
Perilla/Shiso
Rosemary
Sage
Summer savory
Tarragon
Watercress

Table 3.3 Annual/biennial flowers that benefit from pinching out

Ageratum
Alyssum
Calendula
Californian poppy
Candytuft
Cerinthe
Cineraria
Cornflower
Cosmos
Dianthus, sweet william, carnation (not cut-flower varieties)
Marigold
Mignonette
Nasturtium
Petunia
Viola, pansy
Zinnia

can often crush the branch off rather than cutting it. Anvil secateurs do have the advantage of requiring less effort to make the cut, as there is no friction.

By-pass secateurs with a revolving handle are the best option for a precise cut and reduce the effort of cutting by 30%. The revolving handle also helps to prevent callusing as well as fatigue when the tool is used for long periods at a time.

Flower snips are useful for deadheading and cutting small branches and twigs where a larger, heavier secateur may twist smaller stems. Stainless steel snips are also useful in the kitchen! See Figure 3.8 for a selection of basic secateurs.

> If you are left-handed do not attempt to use right-handed secateurs. Using unsuited secateurs will damage your hand and your plants.

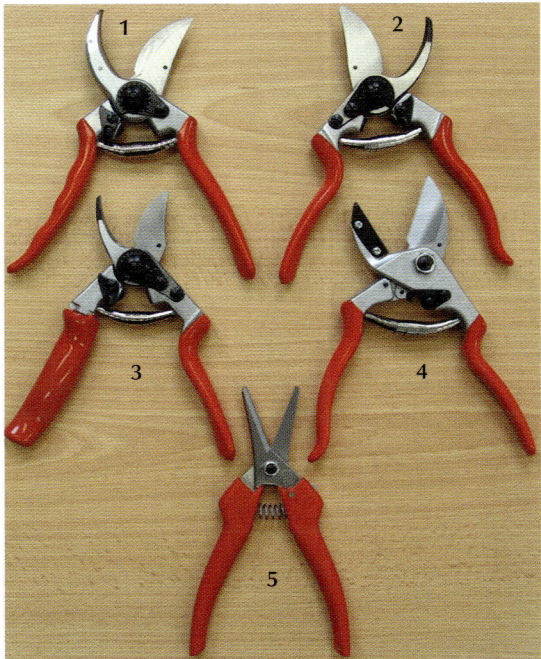

Figure 3.8 A selection of basic secateurs. Right-handed by-pass secateurs (1). Left-handed by-pass secateurs (2). By-pass secateurs with revolving handles (3). Anvil secateurs (4). Flower snips (5).

Maintenance and sharpening

Secateurs/pruners are the most essential of pruning tools and therefore buying good quality tools that can be cleaned, sharpened and have parts replaced is the best investment a pruner can make. Buy the best secateurs you can find and they will most likely be in use for future generations, making them the most economical in the long term. Good quality tools deserve good maintenance. Once you have invested in a good quality pair of secateurs it is important to keep them sharp, clean and well-oiled (see Figure 3.9).

It is necessary to take your secateurs apart to clean and maintain them properly.

Sharpening

The most frequently performed task is of course sharpening your secateurs. Sharp secateurs reduce the effort of pruning and produce a cleaner cut.

Tools

Investing in a sharpener that is designed for secateurs makes the job easy (see Figure 3.10).

Files with embedded industrial diamonds are the most effective, as diamonds, being the hardest stones, will therefore work quickly, last longer and produce fine filings (see Figure 3.11). The Istor blade at the other end of this tool works like a carpenter's plane. It swiftly sharpens the blade, but also wears it down quickly. Use it to rapidly reshape the bevel edge of a worn and blunt blade. Blades with indentations or notches should be replaced; they have had their life of use and abuse and are not worth continuing with.

Technique

First, grasp the secateurs as shown in Figure 3.12 so that they are comfortably secure. The perfect angle to sharpen the blade is 23°. This sounds pedantic and almost impossible, but is in fact easy (see Figure 3.13).

Sharpen the entire length of the blade even if only part of it is blunt. Using a circular motion, apply enough pressure to feel some resistance, until the blade is shiny the whole length of the blade's bevel (see Figure 3.14).

You can now burnish the blade to harden its cutting edge by compression. Burnishing your newly sharpened edge will preserve its

Figure 3.9 Maintaining your secateurs. First remove the spring (a). Loosen the screw that controls the central nut and bolt (b). Remove the nut from the central bolt (c). Take the secateurs apart (d). Clean and grease all the parts as required (e). Reassemble the secateurs (f). Before reinserting the spring, adjust the central nut so that the blade falls naturally to run against the by-pass blade for two-thirds of the blade's length (g). Re-insert the spring and check that the secateurs run smoothly. Adjust the central nut as required (h).

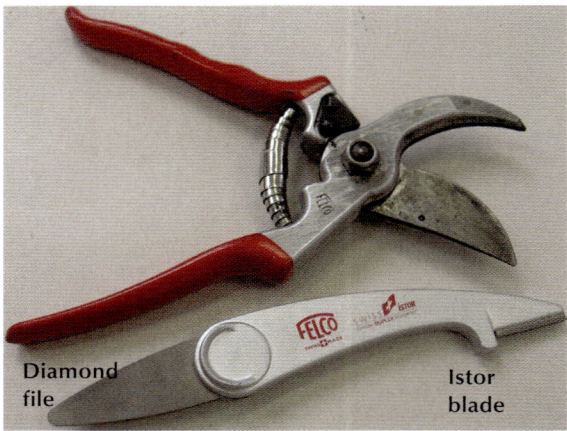

Diamond file

Istor blade

Figure 3.10 The best sharpeners are shaped to allow easy access to all parts of the blade.

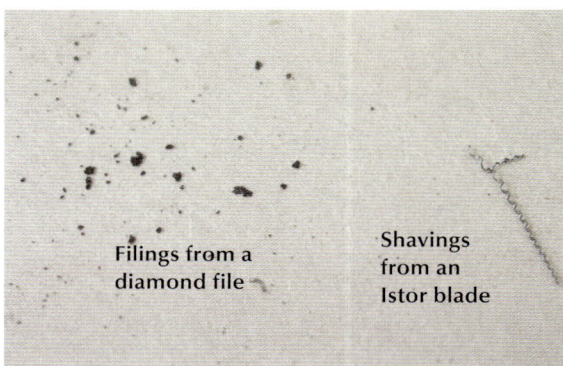

Filings from a diamond file

Shavings from an Istor blade

Figure 3.11 Files with embedded industrial diamonds are very effective, and produce fine filings (left). The Istor blade at the other end of this tool works like a carpenter's plane. It is used to rapidly reshape the edge of a worn and blunt blade (right).

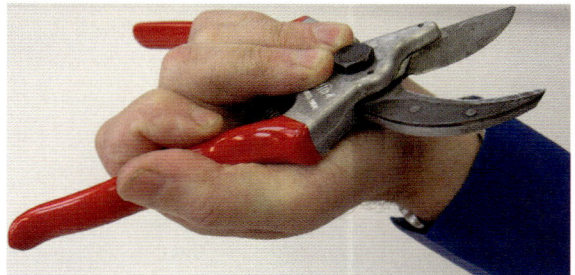

Figure 3.12 A good secure grasp in preparation for sharpening.

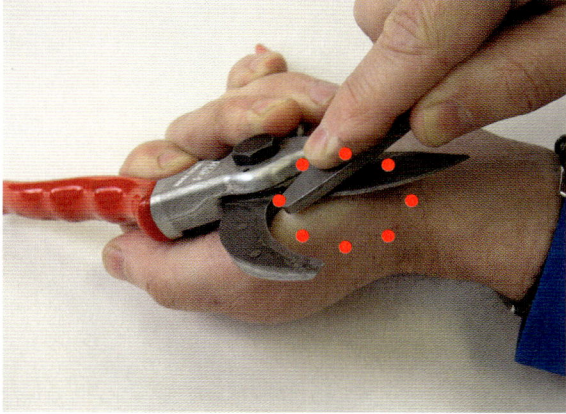

Figure 3.13 The perfect angle to sharpen the blade is 23°. Hold the sharpener at a 90° (right angle) to the blade (top) and then tilt it to half of this angle, that is, 45° (middle). Finally, halve this angle again and you reach the optimum 23° angle for sharpening (bottom).

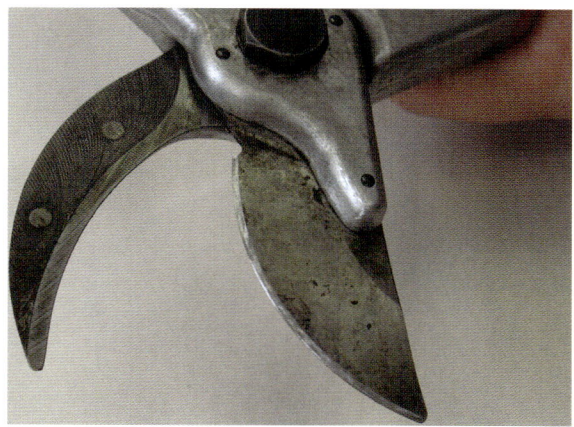

Figure 3.14 When the blade is shiny the whole length of the blade's bevel it should be evenly sharp.

sharpness for longer. Hold your sharpener like a potato peeler with a narrow edge against the blade. Increase the angle of the sharpener so that it is in contact with the very rim of the cutting edge. Press down on the full length of the cutting edge, applying about 5 kilos of pressure, which is about moderate strength (see Figure 3.15).

Turn the secateurs over and remove the burr (or lumpy bit of metal) on the flat side of the blade by running the sharpener at an angle of 2° to 3° along the blade (see Figure 3.16).

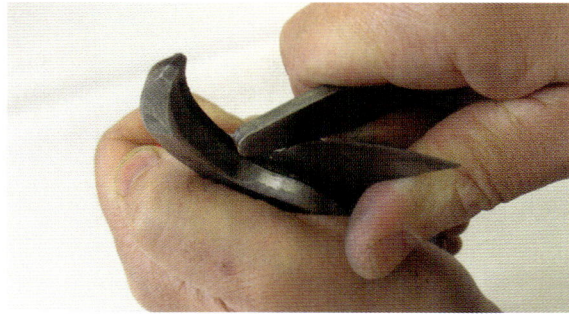

Figure 3.15 Burnish your blade to harden the cutting edge.

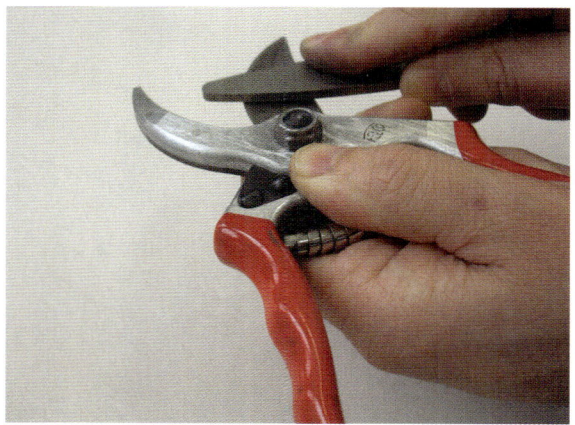

Figure 3.16 Make sure you remove the burr on the flat side of the blade.

Figure 3.17 The final test for a perfect cutting blade.

Test your secateurs by slicing up a blade of grass right along the length of the blade (see Figure 3.17). If the blade fails to do this after correct sharpening, the secateurs central bolt may need to be tightened.

How to cut
Technique
Now that your secateurs are sharp you are ready to cut. Ideally the cut should be made on an angle (to shed water) with the blade next to the bud you are cutting towards (see Figures 3.18 and 3.19).

Figure 3.18 The blade should be next to the bud being cut to and on an angle.

Figure 3.19 The bottom of the cut should not go below the bottom of the bud. If it does the bud may dry out and die.

Leaving a stub above the bud is also detrimental. As most growth in plants occurs from active buds, a piece of blank stem will simply die and possibly be a site for disease to enter (see Figure 3.20).

Commercial orchards and vineyards cannot afford such painstaking work, yet the trees and shrubs they cultivate have to produce the maximum of quality crop for the farmers to stay viable. It makes sense for them to reduce labour costs where they can; the ideal is good but not always necessary. On the other hand the home gardener will not need to prune hectares of trees, so a good hand-pruning technique will result in effective results. Just remember you don't have to be too exact.

Figure 3.20 Leaving a stub above the bud will almost certainly result in a dead piece of stem; a vulnerable spot for disease to enter.

You can also make many small cuts in one go by using hedge clippers (mechanical or otherwise) or with chainsaws. Hedging plants can tolerate different treatment (see page 85).

Medium branches up to 35–45 mm diameter

Tools

Loppers are the ideal tool for medium branches. The long handles give extra leverage and also extra reach to access higher branches. Although you may be tempted to use the secateurs, using loppers for this size branch is safer for you, your tools and your plant. The more robust loppers prevent the branch twisting. Swivelling or twisting branches can result in an uneven cut or a sudden slip that can injure the pruner.

Loppers come in various sizes (see Figure 3.21). The longer the handles, the more

Figure 3.21 A range of loppers that can cut branches up to 35 mm in diameter. Generally the longer the handles the less strain on the pruner. The pair with the longest handles can cut branches up to 45 mm in diameter.

leverage the pruner has, thus reducing muscular strain. Long handles can make working in tight spaces extremely difficult. Those with shorter handles are ideal for working on plants where the branches grow closely together such as roses.

Large branches 35 mm diameter and above
Tools

Pruning saws are the implement of choice for these branches. Those that cut on the pull stroke only are easy to use in tight spaces where branches are growing closely together.

The folding pruning saw is the safest as it can be slipped into a back pocket while not in use. It also means there is less temptation to use the secateurs rather than the saw! A folding pruning saw is always accessible in a back pocket.

Longer pruning saws can cut branches up to 80 mm in diameter. Those that come with a scabbard and have a belt loop attached are the safest to use. Pruning saws are extremely sharp and should never be carried around the garden unless sheathed (see Figure 3.22).

Figure 3.22 Smaller folding saws are easy to carry around safely and can cut branches up to 50 mm in diameter (1). Longer pruning saws that come with a scabbard and belt loop are ideal for branches up to 80 mm in diameter (2).

Where there is more room to maneuver, a bow saw is ideal. It cuts on both the push and pull stroke making light work on large branches. Care must be taken to ensure that the blade is kept tight at all times, as twisting and lack of tension will cause injury to both plant and pruner.

Technique

First, undercut the branch a good hand span from the intended final cut. This prevents the branch removing a large proportion of bark as it falls. Remove the whole branch about 75 mm from the undercut and then make your final cut just beyond the branch collar or ridge (see Figure 3.23).

No plant needs the sort of wound that strips away the cambium layer and leaves a huge

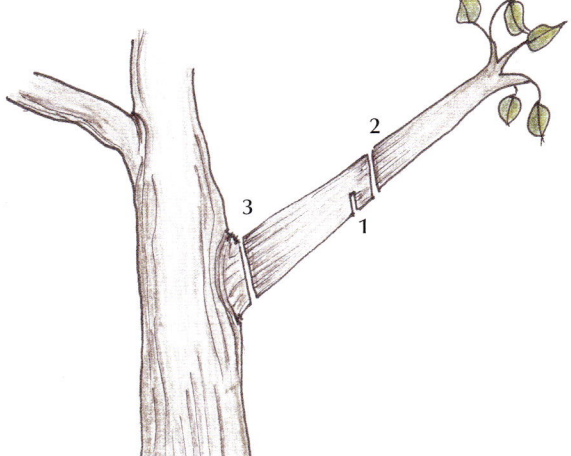

Figure 3.23 Undercut the branch a good hand span from the intended final cut (1). This prevents the branch removing a large proportion of bark as it falls. Remove the whole branch about 75 mm from the undercut (2). Make your final cut just beyond the branch collar/ridge (3).

Figure 3.24 This is what happens when you don't undercut a large branch.

scar, with a large area open to disease. Always undercut the branch before removing it completely (see Figure 3.24).

The best place to cut

Once the bulk of the branch has been removed, the pruner can turn their attention to where to make their major final cut.

All large woody plants have a branch collar or ridge that is made up of a mass of cells that will spring into action to overgrow and protect the wound. This collar or ridge must not be damaged and can be most effective in its healing role when the cut is placed just beyond it. It is usually the most convenient place to cut for hand-pruning as this is the place on the branch where it narrows away from the trunk.

Notching (semi-cincturing) and scoring

This style of pruning is more akin to wounding rather than actually removing plant parts. Wounding produces growth in

Figure 3.25 A range of branch collar ridges indicating where to cut. Choose the point where the top of the branch dips before it meets the branch collar. This point will vary slightly in every species of plant.

plants, something to remember in our own lives! Just as removing a branch at the correct point stimulates the branch collar to heal over a wound, notching and scoring can produce growth on an existing stem. It can initiate new branches, stimulate the growth of more flowering/fruiting wood or in some cases promote fruitfulness throughout the whole tree.

Tools

Use a bluntish knife (yes, blunt) to cut away a piece of bark, or a saw can be used just drawn across the section of stem above the bud. A saw will leave a ragged wound that will be slower to heal/grow over. The blunt knife will also produce a wound slow to heal. A slow healing wound is an advantage in this situation, as the longer the wound takes to heal the more effective is the interruption of the cambium's flow. On really young wood, a fingernail will do the job.

Technique

This technique works best on reasonably young wood up to one year old. If you wish to activate an existing bud into growth on a long stem, cut away a strip of bark or cut with a saw (deep enough to cut the cambium) just above the bud from which you want the new branch to spring (see Figure 3.26).

As the apical bud (see page 3) produces hormones that inhibit the growth of buds below it, removing the cambium at this point interrupts the flow of these hormones (see page 2) allowing the bud to develop into a stem. Removing a strip of bark 3 to 5 mm wide that extends just beyond the width of the

Figure 3.26 This small strip is all that is required to interrupt the flow of the cambium (top). The growth point is fed (bottom) by the phloem without the inhibiting auxin from the apical bud. See page 2.

bud and about the same distance from the top of the bud is all that is required. Alternatively, a saw cut or a fingernail will be sufficient to interrupt the flow of the cambium. As the growth-inhibiting hormone is diverted around the bud, the movement of the nutrient-rich phloem (see pages 1 and 2) feeds the growth point.

Conversely, a notch made just below a bud or branching stem will slow the growth from this point. This can be useful to reduce the growth of an over-vigorous side shoot.

Figure 3.27 A rose stem with a bud to be removed. Cut any thorns so you can get a good grip (a). Using a sharp knife, slice through the cambium layer to the wood above and below the bud (b). Hold the stem firmly and slide the knife carefully under the bud (c). The bud is removed (d).

Disbudding

Sometimes the removal of buds from a stem or root system is required. This technique is used to remove buds from the stem of a standard plant or to eliminate the possibility of suckers from a root system. It aims to remove the meristem that will give rise to another stem or sucker (see Figures 3.27).

Spiralling

A single cut spiralling right down the trunk can stimulate flowering in trees.

This is a technique for the brave, to be used only on overly vigorous but unproductive plants; that is, a plant refusing to flower/fruit.

Tool

A pruning saw.

Technique

Draw a saw down the trunk in a wide spiral (see Figure 3.28).

This wounding puts stress on the plant by interrupting the flow of the cambium throughout the plant. This stress will shock the plant into a mad reproductive urge resulting in flowers and then fruit. Wound your plant about four to three weeks before the plant breaks into bud for deciduous plants, or four to three weeks before it is due to flower for evergreens.

Figure 3.28 This spiral wounding will slow vegetative growth and promote flowering/fruit.

Root pruning

Pruning roots can be tricky. Suffice it to say that pruning roots you can see is much safer than tinkering with a root system that is in the ground.

Root pruning is a technique used to dwarf plants. Removing roots from a plant will decrease the amount of water and nutrients available to the plant, and as we all know, the less we eat and drink the smaller we get!

Bare-rooted plants often have broken and smashed roots that should be removed to reduce the likelihood of infection.

Which roots to prune?

There are a few general types of roots that plants have during their lifecycle.

All seedlings start with a seed root (radicle) that forms the tap root. This rapidly growing root anchors the plant in the ground so that the newly germinated seedling is not washed or blown away. The humble carrot is the obvious example of a strong taproot as well as a food-storing unit for the plant.

Generally the taproot does not persist in long-lived woody plants except some plants adapted to desert conditions. It soon branches to form secondary roots to further exploit the available water and nutrients where there is still plenty of oxygen in the soil. All terrestrial plants need oxygen at their roots, which is why a root system will only extend as deep as the air can penetrate the soil profile. The secondary roots branch again to form more fibrous roots.

At the tip of *all* of these roots there are the roots hairs (see page 2). The last few millimeters of each root is the part responsible for drawing water and nutrients from the soil. These root hairs are easily damaged but regenerate quickly in favourable conditions. It is therefore logical that the more fibrous the root system (the more roots there are) the more root hairs will be produced, thus creating a larger surface area from which to draw sustenance.

Root pruning to transplant large trees and shrubs

When you wish to move large specimens in the garden, it is best to plan ahead. If you plan to move the plant in late winter, it is wise to prepare the plant the previous spring.

Tools

Use a sharp spade or axe.

Technique

Identify where the drip line is (see Figure 3.29). The drip line is the area just beneath the outer edge of the plants canopy where rain drips off the leaves to the ground. This region has the highest density of water and nutrient absorbing roots hairs. Therefore, this is the region where watering and manuring is most beneficial. It is also the place to start cutting roots in preparation for transplantation. Suckers can occur when roots are damaged or the plant is stressed.

Dig around the drip line to a spade's depth and about 15 cm wide. This effectively cuts the main feeding roots and promotes root branching between the cut and the tree/shrub trunk. Pack this space with well-rotted compost so that the damaged root ends have

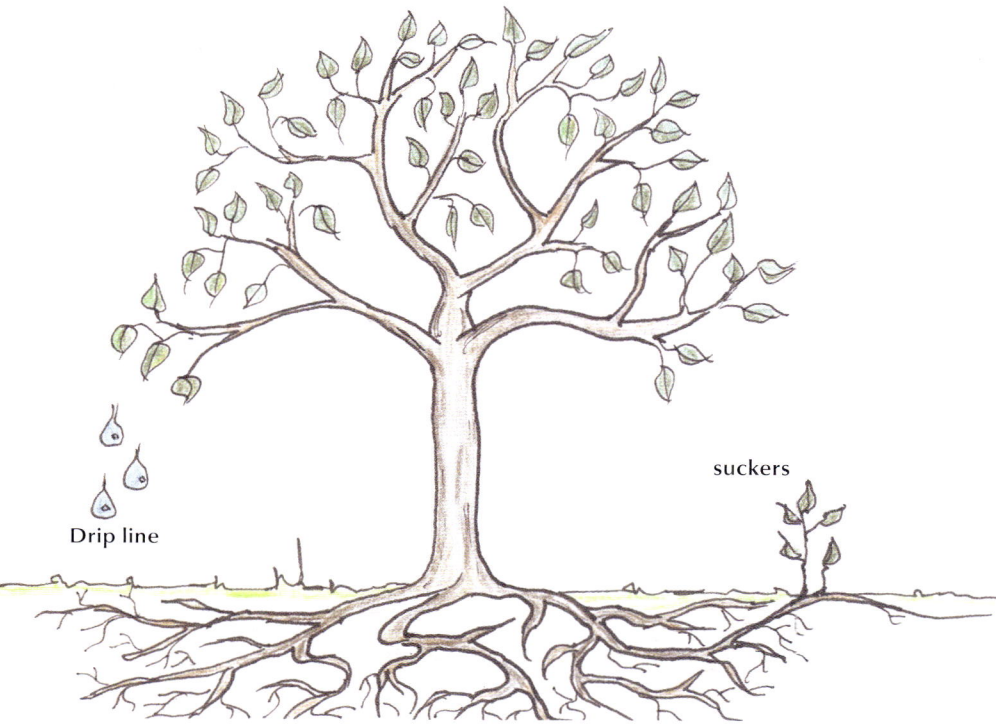

Figure 3.29 The drip line.

plenty of nutrients to regenerate. The following season when the plant is moved, use this trench to define the outside edge from which to commence digging the plant out. There will be a mass of fibrous feeding roots just inside this edge to aid in the successful re-establishment of the plant.

Root pruning for dwarfing

In-ground root pruning is a risky affair. Some orchardists run a tractor blade down the middle of their rows in the well-founded belief that by removing part of the trees root system, the supply of water and nutrients will be lessened resulting in a smaller growing plant. The same effect is achieved by digging with a sharp spade around a plant just within the drip line; however, applying this technique to plants in the ground is a risky business that can lead to a multitude of problems.

Cutting roots in this manner is a hit and miss affair. There is no telling what sort of root you will be cutting. As all wounding creates growth, there is a risk of encouraging suckers. Vigorous roots when wounded and exposed to light can rapidly form tenacious suckering growth that can be a nightmare to control (see Figure 3.29).

If you really want to dwarf your plants in the ground, summer pruning is safer and more effective.

Never try in-ground root pruning on:

- Poplar, *Populus* species
- Bay trees, *Laurus nobilis*
- Elm, *Ulmus*
- *Gledistsia* species
- Figs, *Ficus*
- Apple, *Malus*
- Olive, *Olea europa*
- *Robinia* species
- Quince, *Cydonia* species
- Hawthorn, *Crataegus* species
- Willow, *Salix* species

However, if you are after a thicket/multi-stemmed landscape look, injuring the larger surface roots can be very effective. It is a technique that works well in recreated naturalistic plantings. Many species of *Acacia* (Wattle), *Melaleuca* and *Rosa rugosa* respond and look good with this treatment.

Suckers and how to deal with them

With or without root pruning, suckers occur. A sucker is a plant part that arises from the root system often at some distance from the main trunk. They arise often as a result of damaged surface roots (usually mower damage) as a result of digging near the tree or stressful climatic conditions (see Figure 3.30).

Some species are more prone to this behaviour than others (see the list above). Plants that have been grafted to a rootstock such as many roses or fruit trees will also sucker. In their case the growth that arises from their root system will often look different to the growth above the graft. Suckering can be the result of root damage or of severe stress (drought, lack of nutrients or

Figure 3.30 Suckers can occur away from the trunk. Often the leaves will look different to those on the rest of the tree above the graft. Scratch the soil away to locate where it is connected to the root (a). Always pull, rip or wrench off the sucker so that its growth point is removed from the parent root. **NEVER** cut it off as this will leave plant tissue that will resprout (b). When the sucker is ripped off it will take the growth point or meristem from which it originated (c). Be vigilant in spring and autumn and remove the sucker as soon as you notice it. The younger the growth the better.

Figure 3.31 The pot-bound plant will have taken on the shape of its pot (a). Wash away the soil from the roots so it is easy to see. Cut away the thicker roots saving as many fibrous roots as possible (b).

unsuitable soil conditions) where the vigorous rootstock outgrows the weaker cultivar grafted onto it.

Suckers are best tackled when they are young; the greener the wood the better. If necessary, dig down until you can see where the sucker is attached to the root. Grasp the sucker firmly and *rip* it off its parent root. *Never* cut suckers off; it leaves part of the bud the sucker originated from, so it just resprouts. The idea is to remove the meristem/bud/node entirely so that the sucker will not regrow. Unfortunately, once this suckering habit has started it is hard to stop. Keep your plants healthy and try to be vigilant in removing the suckers as soon as they are noticed. Be on the look-out, especially in spring and early autumn.

Root pruning potted plants

Bonsai is the classic illustration of how plants are kept small by pruning their roots. This is acceptable when you can remove a plant from its pot, wash the roots and get a good look at what needs to be done (see Figure 3.31). Root pruning is a useful tool when growing fruit trees in pots, especially those that are not grown on dwarfing rootstocks (see page 155), like figs.

Once the soil has been washed away it is easy to see the large roots that have few root hairs and the fine fibrous ones that will support many root hairs. As the root hairs will be severely damaged, don't allow the plant to dry out once repotted. By keeping your plant moist, not wet, the translocation of water and nutrients will enable swift recovery.

Large thick roots are really just long plumbing appendages, useful for stretching out in search of water and nutrients at a distance when the plant has a free run of the ground. However, they are not necessary in a pot where the root zone is so confined and the plant needs as many fine roots bearing root hairs as possible, and the supply of water and nutrients is in the control of the gardener.

4
ORNAMENTAL PLANTS

Trees, shrubs, variegated plants, herbaceous perennials, grasses and tufty plants

Pruning ornamental plants is a matter of deciding what makes them ornamental and then enhancing that feature. Is it their flowers, their stems and bark, their fruits or their foliage? Whatever pruning you do, nothing beats good plant selection. Consider your space, aspect, local climate and the shape you want your plant to become. There is no substitute for a qualified horticulturalist, experienced gardener or good research for these decisions. The pruner, no matter how skilled, will never be able to make an upright growing plant into a spreading or rounded plant. Shrubs that naturally make a rounded relaxed cascading shape, like Spirea or Plumbago, will never make a tidy hedge. It is possible to fit a tall spire like tree under a power line, but is it worth the effort? A tree with a rounded crown can be trained to cope with the situation (see Figures 4.1 and 4.2). *Good plant selection for the site can save on pruning and maximise a plant's natural beauty.* A plant's natural shape can be forced into submission through constant pruning and training; however an ornamental plant is just that, ornamental. It exists in our gardens landscapes due to its inherent beauty.

Pruning can only help a plant be the best it can be. Much like the local beauty salon, the beautician can enhance your best assets, but heavy plastic surgery usually results in the loss of individuality. Who would want a garden full of Barbie dolls? A successful garden or landscape is made up of varied individuals, each complementing its neighbor to present a harmonious whole. Choose your plants for what they can add to your landscape, not for what they can be bludgeoned into. An ornamental plant should be allowed to express its personality, and the job of the

Figure 4.1 This liquidambar has a naturally spire-shaped crown that has to be constantly pruned to keep under the power lines. Despite the autumn foliage, it looks cramped and uncomfortable year round.

49

Figure 4.2 This plane tree has a rounded crown which can be opened up to accommodate the power lines. Walking down the street, it looks perfectly natural.

pruner is to make it the most beautiful it can be.

> Remember, no matter what the plant, keep it healthy and remove dead, diseased or crossed limbs.

Ornamental trees

Trees can be valued for their flowers, their foliage, their bark or their fruits; but they are always valued for what they are: trees. They are meant to be statuesque. By definition, they have a trunk (or multi-trunks) before the business of foliage/flowers begins. Their appeal is the quiet dignity of column and canopy enhanced by seasonal variations. Generally ornamental trees can be left to develop by themselves. Care should be taken in establishing their basic structure in their

formative years (see page 15, 'Choosing the right plant at the nursery', and Figures 2.9 and 2.10 on pages 19 and 20).

Trees straight from the nursery may need some initial adjustment (see Figure 4.3). Pruning at this stage should be extremely gentle as quick establishment in the landscape is aided by maximum possible leaf coverage.

Keep pruning to a minimum by buying well-formed stock from a reputable nursery. If there is no alternative:

1. Remove branches at an angle less than 45° from the trunk (unless you are planting fastigiate/columnula trees such as *Cupressus sempervirens*, pencil pine or *Polpulus nigra*, Lombardy poplar).
2. Remove or cut back branches that have a greater diameter than the branch they spring from and may threaten the central leader (future trunk)
3. Remove branches that are too close to each other on the trunk
4. Select branches that alternate vertically around the trunk to prevent shading.

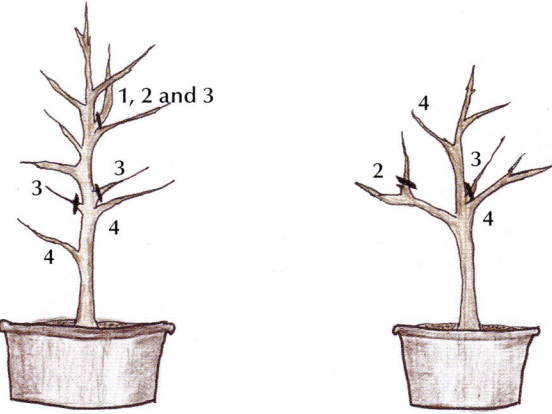

Figure 4.3 Corrective pruning at planting.

Do not shorten back the tips of the branches unless they are damaged. The tips hold an apical bud (see page 3, 'Apical buds') and their job is to extend the growth of the tree, usually at the expense of the axillary buds. This is exactly what is needed to form a good stable branch structure for your young tree, so leave the tips alone. Confine any pruning to promote the central leader and to remove or restrain competing leaders or branches that have a greater diameter than their parent branch. Such branches will eventually destabilise the tree so deal with them before they become a problem. Look at the spacing of the branches. Are they too close together, prohibiting good branch attachment? (See, for example, Figure 2.10). Allow for 30 to 60 cm between branches. Branches growing at an angle to the trunk less than 45° are also unstable, get rid of them now. The only exception to this rule is with fastigiate or very upright slender trees that grow to create a column of foliage like pencil pines or Lombardy poplars. These trees will always have very narrow branch angles.

Try to select and promote branches that alternate around the trunk of the tree vertically. This will ensure that the branches will be balanced and not shade each other (see Figures 1.8 and 1.9). Trees from reputable nurseries should not have these problems so pay that little bit extra for a tree that does not need corrective pruning and will therefore establish quickly.

Keep your tree healthy and the removal of dead, diseased or crossed limbs is enough to ensure its beauty. If you feel that it is in need of renovation or shape modification in the future, see page 102, 'Renovating older trees and shrubs', and Figures 5.23 to 5.33. If your tree has particularly beautiful bark, you may consider making more room under the tree by raising the foliage canopy to show off this feature (see Figure 4.54). Otherwise, just let them be.

In the case of trees with multiple features, such as *Lagerstroemia indica*, or crepe myrtle, with fascinating bark *and* the potential for huge flowers, the pruner must decide which feature they want to emphasise. In order to show off the silky muscled bark cladding the tracery of branches, it is best to just let the plant get on with it without intervention. The plant will still flower if it is kept healthy, but the size of bloom will be smaller.

If, however, the production of large blooms is the priority (some may find them gross), the branches should be cut back by about a third annually after flowering (once it is established) to encourage the new growth that will carry the flowers. This treatment keeps the plant at the large shrub size and therefore it functions as a shrub in the garden design rather than a tree. Trees, apart from maintenance for health, can generally be left to develop by themselves.

Ornamental shrubs

Shrubs are not statuesque except in the case of topiary or dwarf conifers (see page 98). They are largely purveyors of foliage, flowers, stems, and sometimes fruits. The density and magnitude of these features can easily be enhanced by pruning.

Thinning older wood

All shrubs are multi-stemmed and do need their canopy to be renewed over time. Once plants are established and are at a point where the foliage and flowering growth is not as abundant as it is in a new shrub, it is time to assess the stems/trunks. Look at the base of your plant. It is easy to discern between the old stems and the newer when you look at the differing colours of wood on the plant and the differing stem thicknesses. The thicker the stem, the older the wood. Older dark or crusty textured wood is a sign of age with a consequent lack of vigour. This is the wood that should be removed to maintain the shrubs maximum flowering capacity (see Figure 4.4).

Figure 4.5 Remove older stems on this *Phlomis* to encourage strong basal growth.

In Figure 4.5, the *Phlomis* is showing new basal growth. It can be capitalised on by removing older stems.

This theory applies to all shrubs whether the stems originate from the base or from a short trunk. Maintaining younger wood will promote a more vigorous plant (see Figure 4.6).

Figure 4.4 This *Spirea* is in need of renewing. The dark stems are those that are oldest. Simply remove these from the base. Thin, weak stems can be removed also.

Figure 4.6 In quick-growing species, shorten back lanky stems to an actively growing bud or strong side shoot.

Figure 4.7 Note that most new growth is coming from the younger wood. Remove the crusty old wood to make room for new growth.

Complete canopy removal/rejuvenation

Some shrubs can be cut back to stumps, virtually decapitated, and bounce back refreshed. This technique is an alternative to replanting and, despite a short ugly period following the severe pruning, it is very effective in renewing the flowering wood of the plant and reducing its overall size. If the canopy is not removed and it is simply clipped back, you will be creating a hedge out of what was once a graceful informal shrub. As soon as you feel inclined to bring out the hedging shears, pause and consider whether this is what you want your plant to become – a hedge

or topiary. If not, the removal of much of the canopy is the answer. In Figure 4.7, for example, the buddleja has been cut back to stumps. It is reshooting vigorously. Pinch out the tips of these shoots to promote bushiness and the natural relaxed shape of the plant will return.

This procedure need only be attempted every five to seven years depending on the vigor of the species. Some plants respond by producing new suckering growth, others by resprouting from the remaining wood. See Table 4.1 for species that tolerate this treatment. See also 'Coppicing', page 84.

Table 4.1 Shrubs that respond to decapitation for canopy renovation

Acacia boormanii
Banksia marginata
Brugmansia versicolor
Buddleja spp.
Callistemon spp.
Calothamnus quadrifida
Camellia sasanqua and *japonica*
Cestrum spp.
Coprosma spp.
Cornus alba
Cotoneaster spp.
Dodonea viscosa
Escallonia spp.
Grevillea 'White Wings'
Hydrangea species
Melaleuca armillaris
Melaleuca ericifolia
Nerium oleander
Pyracantha spp.
Rosa rugosa (cutting grown)
Salix spp.
Sparmannia africana
Spirea spp.
Symphoricarpos spp.
Telopea specioissima
Thryptomene saxicola
Viburnum odoratissimum
Viburnum opulus

Timing

Complete canopy removal and the removal of old wood should only be carried out just before a season of growth. That is in early spring for those in frost-free climates. In areas prone to frost, wait until all threat has passed before proceeding.

Flowering shrubs

Some shrubs flower just once a year. Other shrubs repeat flower over the season, while some never seem to want to stop flowering, like daisies and French lavender. Flowering shrubs, no matter when or how often they flower, should be pruned with the knowledge of what sort of growth they flower on and the structural strength of the shrub in mind.

Many shrubs need hardly any pruning at all once established. They may require a bit of shaping and shortening back of overly vigorous shoots, but pruning can be limited to a bit of wood thinning and the removal of dead or diseased wood. If the plant is getting overly woody, prune after flowering to a vigorous side shoot pointing in the direction you want the growth to go. You may want to remove spent flowers by deadheading, that is, cutting the flower stem off to the nearest vegetative bud (see Table 4.2).

Table 4.2 Shrubs that need minimal pruning

Camellia
Choisya ternata Mexican orange blossom
Continus Smoke bush
Escallonia
Fothergilla
Gordonia
Hibiscus syriacus Deciduous hibiscus
Loropetalum
Magnolia
Melaleuca spp.
Michelia spp.
Murrya paniculata Orange blossom
Nandina
Nerium oleander Oleander
Peonia Tree peony
Protea
Rhododendron
Teucruim fruticans
Viburnum plicatum

Shrubs that flower once a year

Many shrubs flower on the new growth resulting from last year's growth (see Table 4.3) *Whatever the case, the standard rule of thumb is to prune directly after flowering to an outward-facing bud.* Shrubs that also carry ornamental berries/hips are treated differently (see below). Once your shrub has become established, shortening back vigorous shoots and pruning out spent flowers has the advantage of saving the plant the energy needed to form seed, a metabolically expensive exercise for the plant. The energy saved can then be directed into providing new growth to carry next year's flowers.

It is important to prune shrubs that flower early in the season straight after they have flowered before the new growth has fully developed. Pruning after this time not only means that you put up with dead flowers, but also that the long lanky growth will produce few side shoots to carry next year's flowers. By removing the end of the branch (removing the apical/terminal bud) new growth is forced sideways making your shrub bushier and more floriferous next year (see pages 3–4, 'Apical' and 'Axillary buds', and Figures 1.4 and 1.5). The viburnum in Figure 4.8 illustrates the general principle.

Always cut behind the flowers no matter how tempting the new growth may be. If you don't, the result will be a very straggly affair with few flowers. Plants that put on new vegetative growth almost as soon as the flower opens are the hardest for the new pruner to remove. *Callistemon, Rhododendron* and *Daphne* species are examples. Harden your heart and cut to behind the flower or you will be left with a straggly mess. Another advantage of pruning at or after flowering is that you can cut stems to enjoy indoors, with the added joy that you are helping your plant reach its full flowering potential (see Figure 4.9).

Table 4.3 Shrubs to prune after or during flowering

Ballota
Buddleja spp. Butterfly bush
Callistemon Bottlebrush
Choisya Mexican orange blossom
Cistus spp.
Correa Native fuschia
Crowea
Deuztia
Erica Heath
Eriostemon spp.
Erysimum spp.
Forsythia
Garrya eliptica Tassel bush
Genista Broom
Grevillea Spider flower
Hebe
Kolkwitzia Beauty bush
Lavandula angustifolia English lavender
Lavatera
Leonotis
Leptospermum
Lupinus arboreus
Perovskia atriplicifolia Russian sage
Philadelphus Mock orange
Plectranthus
Plumbago auriculata
Rosmarinus Rosemary
Sambucus Elderberry
Streptosolen jamesonii
Syringa Lilac
Tamarix
Thymus spp. Thyme
Viburnum x burkwoodii
Viburnum opulus Snowball tree
Weigelia florida

Figure 4.8 This viburnum flowers on buds arising from last year's growth. It is easy to see by noting the colour of last year's growth (dark) and the present spring growth (bright green) that has produced the flowers. Cut back after flowering to behind the flowers. On weaker shoots that you want to encourage, cut back to the lowest flower and deadhead (a). If you wish to restrain growth cut to behind the flowers (b).

Shrubs with flowers followed by decorative fruit/berries

Shrubs grown for their flowers and decorative fruits require different treatment from the above. Fruit follow flowers, so if the flowers are removed there will be no fruit display. Any rampant vegetative growth put on in spring can be shortened back so that the

Figure 4.9 For shrubs that produce new vegetative growth almost as soon as the flower has opened, be brave and cut to behind the flower. If you don't, your shrub will be straggly mess.

symmetry of the shrub is maintained, but leave the rest alone unless you want to sacrifice the autumn display.

Any rejuvenating pruning should be done after or during the fruiting display (see Figure 4.10).

Again, this is a good opportunity to cut the fruits for display indoors. The same principle

Figure 4.10 Prune berry plants during or after the fruiting display.

applies in the removal of old wood. Prune back older wood to a strong side shoot heading in the direction you want it to grow in, or remove old stems altogether if there is evidence of strong younger growth clamouring for space. It must be noted that many shrubs that produce autumn fruit can also become environmental weeds. Always choose what species you plant carefully, for example, *Cotoneaster* spp. Your local council will have a list of pest plant species.

Variegated and coloured foliage plants

Shrubs with variegated foliage add a bit of sparkle to the garden scene and can balance a landscape reliant on flowers alone. They act as a foil for the more glamourous members of the landscape and are often the most spectacular plants in winter when there is little else to attract the eye.

Variegated and highly coloured plants are selected chance seedlings or sports and as such are not typical of the species from which they derive. In fact, they are an aberration. Highly

coloured and variegated leaves are actually a less efficient leaf than a plain green one. Variegated leaves have less green parts that absorb light (see page 5) due to the presence of the variegation. For this reason a variegated plant will often revert to the leaf colour of the

Figure 4.12 When variegated plants are grown in low light conditions, the leaves can revert. Ivy may be a shade-tolerant plant, but variegations are not (a). Black and highly coloured leaves are an aberration just like variegations. In the centre of this plant, deprived of light the leaves are green-not the black that has access to the light (b).

Figure 4.11 When variegated plants are stressed, they will often revert to the original green colour of the species. Remove all such parts, just behind the next variegated bud.

original species, especially if it is under stress (see Figure 4.11).

Plants with deeply coloured leaves are similar. Their deep purple or black leaves mask the chloroplasts beneath, inhibiting photosynthesis.

Most often reversion occurs as a result of light deprivation; the plant is reverting to the leaf that will gather the most light from what is available, that is, the wholly green leaf. When they are grown in low light conditions, their handsome leaves may revert to the botanically more practical green (see Figure 4.12).

The best solution is to move the plant to a sunnier spot where it will not be so desperate to clutch at every available sunbeam. Sometimes reversion occurs as the result of an unstable cultivar and no matter where it is in the garden it will produce wholly green leaves. The best solution in all these cases, apart from moving the plant, is to cut out the undesirable foliage immediately from where it originates on the plant. That is, remove the entire branch or stem that is carrying the undesired leaves (see Figure 4.11).

Shrubs that don't stop flowering

It is all very well to say 'prune after flowering', but what of those hardy stalwarts that seem to go on forever? What about the common old daisies and the French lavender?

As most people that value plants love flowers, we are reluctant to interrupt such generosity as a seemingly endless display. But interrupt it we must.

These ground covers and sub-shrubs grow constantly in congenial conditions. They produce new flowering growth often from spring into winter. All the time they are growing larger and larger, covering the spent blooms with a new mantle of fresh flower, while underneath that mantle they are becoming leggier and more structurally unstable by the day. If left to their own devices they will collapse completely, leaving a gaping hole, or at the very least become a sprawling mass of stems toped by a wisp of foliage and the inevitable flower.

Sub-shrubs that behave in this manner (see Table 4.4) should always be prevented from flowering until they have developed a reasonable structure of branches from which to launch their flowers. Pinch out any flower buds for at least the first two to three months after planting (see Figure 3.7). Not only will this help the plant establish, it will build a dense cradle of branches that can be built on. After that you can let them rip – for a little while. After two months of constant flower, it is time to call a halt. Cut back the flowering stems to a strong side shoot, thus encouraging

Table 4.4 Shrubs that don't stop flowering

Ageratum houstonianum
Alyogne huegelii Native hibiscus
Arctotis x hybrida African daisy
Argyranthemum spp. Daisy bush
Brachyscomb multifida
Diascia spp.
Erigeron karvinskianus Seaside daisy
Felicia ammeloides
Gaura lindheimeri
Lavandula dentata French lavender
Lavandula pedunculata Spanish lavender
Penstemon
Salvia mexicana Mexican sage
Verbena x hybrida

branching lower down in the bush. Shorten back long stems to strong growth (see Figure 4.13).

It is a good idea to pinch out the tip of all new growth at this point (see Figure 3.7). This process can be repeated every few months.

However, if you are facing an almost exhausted shrub of this kind, there may be very little viable growth to prune to once the flowers are removed. This can be problematic. The removal of so much growth will leave very little leaf for the shrub to survive on, and depending on the season, they may not resprout from old wood. A major prune such as this can be staged. It will look odd, but cut half of the shrub and leave the other half until the pruned section is showing vigorous growth. This sort of operation is best performed in spring after the risk of frost.

Groundcovers that barely stop flowering can be treated more brutally but with the same good effect. They can be grasped by their longer stems and simply sheared off, leaving green stems at the base (see Figure 4.14).

Figure 4.13 Cut long lanky growth back to a strong side shoot (a). Shorten back spent flowering stems to healthy growth (b).

Figure 4.14 Cut down the old stems almost to the ground, making sure there is leafy growth below the cut.

Figure 4.15 Plants with large and long flowering stems can be cut back in the same manner as Figure 4.14. To lessen the aesthetic down time, prune each strong stem back to a side shoot.

Some groundcovers such as *Arctotis* can benefit from gentler handling, not that they mind the rough stuff! Cutting back every single flower stem individually, and pruning to side shoots is a slow process; however, there is less aesthetic down-time from a landscape point of view (see Figure 4.15).

Grasses and tufty plants

This broad heading includes all those plants that only have an apical bud at the end of each leaf, no axillary buds (see page 3, 'Apical buds', and page 4, 'Axillary buds'). Once the tip of the leaf is damaged, it will not regrow. It cannot grow sideways like a woody plant as there are no axillary buds. These plants are all monocots, or plants that have only one seed leaf and are characterised by having parallel veining in their leaf or stems. Palms and bamboos are also part of this botanical group but their pruning needs are minimal, requiring only the removal of dead leaves and/or fruits. Bamboo may require the canes thinned by removing the entire cane at ground level. If the cane is pruned below the side leaves, it will not regrow.

Grass-like species and many evergreen bulbous plants require more attention. These plants have a different source of growth than woody trees and shrubs.

Grassy leaves emerge from an *intercalary* meristem. This is a point of growth above the root system that gives rise to the actual leaf blade, reeling out more leaf even if the tip of the leaf is damaged (see Figure 4.16).

It is this structure that allows for animals to graze or humans to mow and the plant to

Figure 4.16 Grasses are made up of a fibrous root system, a hollow stem leading to an intercalary meristem, then the blade.

regenerate quickly. When the intercalary meristem is damaged by overgrazing, mowing the lawn too close or cutting back an ornamental grass too severely, the plant will die as there is no point from which the leaf blades can be replaced. Bulbous plants with a similar growth habit such as *Knifophia* are horticulturally treated in the same manner, though many of these plants can regenerate from their bulbous root systems. The same theory applies to stoloniferous grasses such as kikuyu, buffalo and couch.

Table 4.5 Evergreen and *deciduous grasses.

Arundo spp.
Arundinaria spp.
*Calamagrostis spp.
Carex spp.
Cortaderia spp.
Cymbopogon spp.
Desschampsia spp.
Danthonia spp.
Erianthus ravennae
*Miscanthus spp.
Poa spp.

*This can be dependent on climate.

Evergreen ornamental grasses

Evergreen grasses replace their leaf canopy throughout the season leading to a build up of old leaf blades and flowering stems (see Table 4.5). Once the plant has established, usually in its second year, it can start to look rather drab with the old leaf growth crowding and blocking the new. Pruning to rejuvenate the canopy is simple. Grab the hedge shears or the secateurs and remove all the leaf growth *above* the intercalary meristem. This point of growth is relatively easy to locate by looking at the base of your plant (see Figure 4.17).

If the growth is pruned off below the meristem it will die; regrowth is unlikely. For those who want to maintain the soft graceful outline of grasses rather than the hedgehog look, comb through the leaves with you hands (see Figure 4.19).

This is time-consuming but the results are so superior to the shearing technique; as to make it worth your time, unless you have vast swards to attend to! Many Australian native grasses can benefit from cool burning. This effectively removes the dead leaf blades and

Figure 4.17 The intercalary meristem is located where the stem joins the leaf blade. This is the point where the blade grows from even when the tip of the blade is cut. Never cut below this point.

the fallen seed is smoked, promoting germination and ultimately a denser sward (see Figure 4.21).

If the grass is getting very congested, it may be time to lift and divide the clump. Some

Figure 4.18 Grasses can regenerate from tillers, shoots that develop from the base of the plant. This is why grasses are 'tufty'. They may not be strong enough to regenerate the plant if the intercalary meristem is cut.

Figure 4.19 For a more natural look, remove old leaves by combing through them with your fingers.

grasses produce stolons or side stems that can give rise to a new grass plant. These are the toughest of grasses and extremely difficult to kill. These grasses can be cut to the ground safely (see Figure 4.20).

Deciduous ornamental grasses

Deciduous grasses (see Table 4.5) approach the season in an orderly fashion. Spring announces the new leaf growth followed by the development of flower stems and then flowers in autumn. At the onset of winter all growth dies, often providing effective autumn colour and then a straw-like winter sculpture, the perfect foil for frost. These withered stems need to be pruned to the ground. Care must be taken to do this before the new spring growth arrives leaving a significant gap in the

planting. If you enjoy the winter foliage for too long, the old stems must be cut out individually so as not to remove the new growth at the same time; a time-consuming job. In extreme climates where there is heavy frost and snow, the foliage can be left longer as it protects the root system below and the new growth won't appear until later due to the low temperatures.

Evergreen, bulbous strappy leaved plants

These plants (see Table 4.6) need very little attention apart from the removal of spent flower stems and lifting and dividing when they become overcrowded. However, others can reach the end of the season looking tattered. The *Knifophia* (Figure 4.22) has a

Figure 4.20 Grasses grown from stolons can easily regenerate no matter how closely they are shaved. The stolon acts as a food storage unit, and can create a new plant at every 'joint 'on the stolon.

Figure 4.21 This grass has come to the end of its aesthetic life, but it can be rejuvenated by pruning (a). Burn off the dead leaves. If the weather has been damp a bit of newspaper will help it along (b). Make sure you have water available so it doesn't get out of control (c). The intercalary meristem will survive and will produce fresh new blades. Any seeds will also have space to germinate (d).

mass of dead leaves that not only look ugly but need to be removed so that there is room for new growth. It is easily pulled away by combing the foliage with your fingers or cutting off the damaged portion of leaf blade.

All such plants should be deadheaded as their flowers fade by cutting off the flower stem as close to the centre of the plant as possible. At clean-up time, the remains of old flower stems are easily removed (see Figure 4.23).

Table 4.6 Evergreen, bulbous strappy leaved plants

Agapanthus African Lily
Aphelia spp.
Arthropodium spp.
Chorizandra spp.
Cyperus spp.
Dietes spp.
Dierama spp. Fairy fishing rods
Dianella spp.
Diplarrena spp.
Gahnia spp.
Isolepsis spp.
Juncus spp. Rush
Liriope spp.
Lomandra spp.
Libertia spp.
Omphiopogon spp. Mondo grass
Patersonia spp.
Sisyrinchium spp.

Figure 4.23 The remains of old stems can be pulled away to make room for new growth.

Figure 4.22 Dead leaves need to be removed to restore this *Knifophia's* beauty. Pull or comb them out with your fingers.

Figure 4.24 This plant may look slightly naked but now it has the space to reclothe itself with fresh growth.

The result is a plant open to moisture and sun that can get on with the job of being beautiful (see Figure 4.24).

Herbaceous perennials

The most beloved of all flower gardeners, herbaceous perennials are quick growing small plants that do not form woody tissue yet persist in the landscape for at least three years. Depending on species and climate, they may be evergreen or deciduous and are grown for their flowers and/or foliage. Many plants will die back to a persisting root, crown or tuber that will recreate the whole plant again in spring.

Many flowering species require deadheading throughout the season. This is the removal of any spent flowers so that new flowers can be produced. Simply cut the

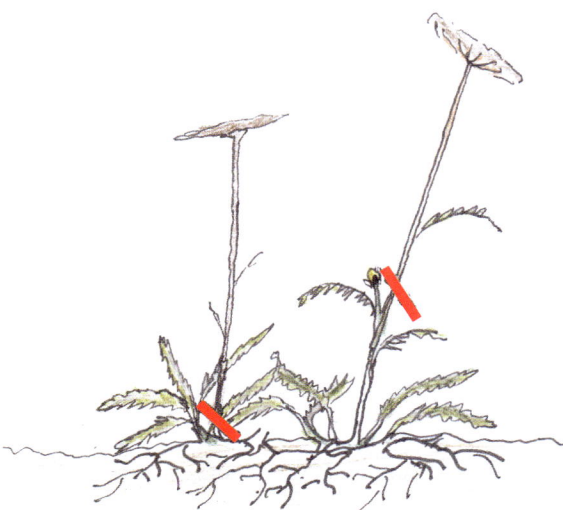

Figure 4.25 This *Achillea* can be cut back to a viable bud on the flowering stem or to the ground if there are no side buds.

Table 4.7 Plants to deadhead through the season

#*Alcea*, Hollyhock (rub off spent flowers)
**Anthemis* spp.
‡*Bergenia*
Centranthus ruber, Valerian
Cosmos atrosanguineum
‡*Crambe scaberima*
#*Cynara* spp., Artichoke, thistle
#*Delphinium* spp.
†*Epimedium* spp.
‡*Euphorbia* spp.
Gaillardia spp.
†*Heleborus* spp.
†*Heuchera* spp.
#‡*Leucanthemum superbum*
Limonium spp.
Linaria spp
‡*Pelargonium* spp.
Penstemon spp.
Phlox paniculata
†*Pulmonaria* Lungwort
#*Pygelius* spp.
‡*Romneya coulteri*
#*Rudbeckia*
#*Salvia ambigens*
Salvia chamaedroides
#*Salvia farinacea*
Tanacetum spp.
#*Thalictrum*
Trachelium caeruleum
#*Veronica*

When the flowering season is over
* Cut down in the dormant season to a base of sturdy branchlets.
# Cut to the ground completely, new growth will come (or has arrived) from the base.
† Remove old leaves before the next growth season.
** Plants that may be cut back by two-thirds and will rebloom in the same season given good growing conditions.
‡ Evergreen species should retain their leaves over the dormant season; only the flower stems are removed.

flower stem off at its base. This will either be at the base of the plant (Figure 4.25) or back to the next bud, node or growth point (see Table 4.7).

Figure 4.26 This dahlia can be deadheaded through the season by removing spent flowers from their point of origin. The main stems will die off in winter. They can be cut to the ground in autumn or removed when the tubers can be divided in early spring.

Figure 4.27 Oreganos and mints can have their flowering stems removed completely as soon as they are no longer an aesthetic asset. Depending on climate there may or may not be new growth at the base of the plant. In some areas it may not emerge until the next growing season.

Other perennials will only flower once and many can be left to hold their old flower stems through autumn into winter, it depends on the structure of the dead flower heads; some plants do look great dead! Others you may want to remove immediately after the flowers have faded and the leaves are wilting as with mints and oregano (see Figure 4.27 and Table 4.8).

There is an advantage in delaying this work. By early spring the old stems will be crisp and light to remove, and in cold climates they would have insulated the soil against frost.

Table 4.8 Plants to be cut down all at once

#*Acanthus* spp.
#*Achillea* spp. Yarrow
‡*Anemone x hybrida*
‡*Anigozanthos* Kangaroo paw
#*Aquilegia* Granny's bonnets
#*Aster*
*Coreopsis
#*Geranium* spp.
#*Gypsophila* spp.
#*Helianthus tuberosum* Jerusalem artichoke
#*Lobelia speciosa*
‡*Lychnis*
#*Mentha* spp. Mints
**Nepeta spp. Catmint
#*Origanum* spp. Oregano
#*Sedum*
#*Setaria palmifolia* Buddha grass
#*Solidago* Golden rod
‡*Stachys byzantina* Lamb's ears

When the flowering season is over
* Cut down in the dormant season to a base of sturdy branchlets.
# Cut to the ground completely, new growth will come (or has arrived) from the base.
** Plants that may be cut back by two-thirds and will rebloom in the same season given good growing conditions.
‡ Evergreen species should retain their leaves over the dormant season; only the flower stems are removed.

Figure 4.28 *Gaura* is a notorious 'flopper'. To prevent this and avoid staking, trim back the early spring growth (a). This provides a stronger framework from which to flower (b). At the end of flowering, prune back the flowering stems to the ground. Depending on your climate there may or may not be fresh growth at the base of the plant (c). In some areas it may not emerge until the next growing season.

Tall herbaceous perennials (see Table 4.9) often need staking – an ugly alternative to their complete collapse. This can be prevented. When the plant has gained about a third of its ultimate height, lightly trim the new growth to form a strong structure from which to flower (see Figure 4.28).

Table 4.9 Plants that benefit from mid spring pruning

*Caryopteris clandonensis
#*Eupatorium purpureum*
*Gaura lindheimeri
‡*Heliotrope arborescens*
#*Monarda spp*, Bergamot
#*Solidago*, Golden rod

When the flowering season is over
* Cut down in the dormant season to a base of sturdy branchlets.
# Cut to the ground completely, new growth will come (or has arrived) from the base.
‡ Evergreen species should retain their leaves over the dormant season; only the flower stems are removed.

Roses

There are few humans that do not respond positively to the rose. Roses are so revered that many seem to think that they are difficult to grow. They require constant 'work' and someone with almost mystical powers to grow them to perfection. After all, how can something so beautiful be anything but difficult to attain? Yet, when it comes down to it, they are flowering shrubs much like all the others. In fact, they are usually hardier and more resilient than many shrubs.

The rose, despite her delicate image, is the child of thugs. To be more polite, roses share their family tree with blackberries and the species roses can often grow into a magnificent bramble to equal those around Sleeping Beauty's castle. They may or may not need taming to fit into our landscapes, but one thing is certain, their beauty comes easily.

As roses have been cultivated for thousands of years, humans have modified and bred them more than any other ornamental plant. Consequently, there are not only the species roses, but many styles of rose that require simple but different pruning techniques. A few are common to them all.

All roses crave sun and air so plant them where they will receive a good six hours of sunshine with good air circulation. Climbers planted against walls should have their main stems placed evenly in a fan shape with about 45 cm or more between them. The length from the tip of your fingers to your elbow is a good rough guide. Humid conditions and roses do not mix so always make sure that the centre of your bush is open and that the stems of climbers do not grow too closely together. The more air circulation in the plant the less disease there will be.

In the following text, the terms 'old wood', 'young wood' and 'twiggy weak growth' are used. See Figure 4.29 for their general appearance.

When pruning any rose always cut to an outward-facing bud to keep the centre of the plant open (see Figure 3.18 and 'How to cut', page 33).

Figure 4.29 Three types of wood are referred to when pruning. Old wood (1). Twiggy weak growth (2). Young wood (3).

Note: The tables in this section list just a small sample of roses available. Those listed as repeat flowering may produce repeated generous flushes of flower under the right conditions or merely produce a few odd blooms.

Pruning at planting

Generally we want roses to be clothed from head to foot in foliage and flower. In order to keep the growth active on the base of the plant, roses should be pruned hard at planting. This applies also to standard and weeping roses that effectually 'start' some distance from the ground as they are grafted onto a long interstock that forms the 'trunk' of the rose.

Usually roses are planted at one to two years old; a young plant. At planting, cut the stems down to three to four buds above the graft (if

Figure 4.30 When planting any young rose always cut the stems back to three or four strong buds; the uppermost outward-facing. Grafted bush or climbing rose (1). Ungrafted bush or climbing rose (2). Weeping or standard roses (3).

it is grafted) or from the ground if it is not (see Figure 4.30).

This applies to species, shrubs ancient and modern, climbers, ramblers, miniatures and carpet roses. If this step is missed, the new growth in the spring will be sprouting at some distance from the base and it may be difficult to induce growth low on the plant in the future. It also provides a good structural base for a long lived and productive plant.

Pruning time

Roses are usually planted when they are dormant and also much cheaper! If you live in an area prone to heavy frost, delay any pruning to early spring even if you have planted in early winter. The frost will only burn off the new growth stimulated by

pruning. Those in less severe climates can prune as they plant.

Containerised roses are now common so they can be planted at any season; just be sure to prune them back as described above in their first dormant season (see Figure 4.30).

Roses that flower once a season

This category comprises of the old fashioned Gallicas, Centifolias, Albas and Damask roses. It also includes the Scotch roses, species roses and sweetbriars (see Table 4.10).

The latter group requires barely any pruning except some deadheading. Figure 4.31 shows a modern repeat-flowering rose. An old-

Table 4.10 Roses that flower once

Alba roses
Cuisse de Nymph (Maidens Blush)
Felicite Parmentier
Mme Plantier
Pompon Blanc Parfait
R. alba
Centifolia roses (cabbage roses)
Fantin Latour
Juno
Petite Lisette
R. centifolia
Damask roses
Belle Armour
Ispahan
Mme Hardy
Quatre Saisons
Gallica roses
Belle de Crecy
Belle Isis
Cardinal de Richelieu
Duchesse de Montebello
R. gallica
Rosa Mundi
Tuscany
Moss roses
Alfred de Dalmas
Chapeau de Napoleon
Henri Martin
R. centifolia muscosa
White Bath
William Lobb
Multiflora roses
R. multiflora
R. multiflora alba
Rugosa roses
Agnes
Conrad Ferdinand Meyer
FJ Grootendorst
Fimbriata
Frau Dagmar Hartopp
R. rugosa
R. rugosa alba
R. rugosa rubra
Roseraire de l'Hay
Scabrosa
Roses with attractive hips (or heps)
Fritz Nobis
Geranium
R. macrophylla
R. moyesii
R. rugosa
R. × highdownensis
Scabrosa

fashioned rose will have roses on a single short stem. Deadheading roses will encourage more flowers in repeat-flowering roses and save others from forming seed.

The only exceptions are roses grown for their hips (see Table 4.10); however, if they are threatening to engulf the house or have become too overcrowded they can be pruned back very hard with no ill effect.

Climbing roses in this category (after pruning at planting) need only weaving in the stems to their support. Some can be trained through trees. You may wish to do some minimal pruning after flowering. Pruning to rejuvenate these roses is best done just after flowering. Many climbing once-flowered roses such as *Rosa banksiae* can be pruned with a chainsaw back to a main framework, and removing some of the oldest wood to encourage new canes. Do not remove too much old wood as the flowers spring from older (but not ancient) wood. This reinvigorates the plant and it may be safely left to its own devices for a few more years; time enough to express their exuberant personality!

The more sophisticated Gallicas, Moss, Centifolias, Albas, Rugosas and Damask roses are best lightly pruned after they have flowered in much the same way as other shrubs that flower once a year. As always, remove any diseased, dead, damaged or crossed stems. Some of the stems may look a bit twiggy and weak; these can be thinned or removed. If in doubt don't prune. As these roses flower on buds formed in the previous season, don't remove so much wood as to reduce the leaf canopy too much. If you do,

Figure 4.31 Remove the flowered stem back to the next bud or side growth.

Figure 4.32 Shorten side growths by about a third always to an outward-facing bud.

there will be little growth after pruning to hold the next season's flowers (see page 5).

If they are becoming too overcrowded in the centre, or you wish to stimulate young flowering stems, prune in late winter or spring depending on your climate. When the time is right for your climate, cut out completely about a third to a quarter of the oldest stems. Prune the remaining stems back by about a third, always pruning to an outward-facing bud. The side shoots can be shortened by a third (see Figure 4.32).

Of course the next season's display will be sacrificed because much of the previous season's wood will have been removed (see Figure 4.33).

As much as a rose that only flowers once seems stingy to modern eyes, their display is generosity itself. Instead of eking out flowers over a whole season, the shrub or climber will smother itself in bloom. Many also have the most stunning hips or fruits to be enjoyed in the autumn (see Figure 4.34).

Figure 4.33 After removing dead, weak or crossed wood reinvigorate once flowering roses by removing a third to a quarter of the oldest stems. Shorten back long stems and side growths by a third.

Figure 4.34 Many once-flowering roses have ornamental hips; do not deadhead them as you will be removing the autumn display.

By removing dead flowers in summer, the subsequent hips will also disappear. Rejuvenate your bush as you see fit by adopting the same method outlined in Figure 4.33.

> When pruning these kinds of roses, keep in mind the character of the rose you are dealing with. They have their own wild romantic charm. If you want 'neat', plant carpet roses and hybrid teas.

Repeat-flowering roses

These are the roses most of us encounter and they are also divided into a few pruning groups. Generally speaking, hard pruning produces bigger flowers but also less shrub. Unless you are growing roses for the local horticultural show, your rose is also a member of its landscape group as a shrub or a climber. It should be considered as a plant in its own right, not just as a few sticks supporting huge flowers.

These roses can be pruned to flower at a certain time. If you have a special event in the growing season where you want the roses to be at their best, lightly prune the newer growth by about a quarter, six to eight weeks before the event. This is not a guarantee of bloom as it depends upon climatic conditions; water and nutrients can be supplied, but warm sunny weather cannot be relied upon.

Large-flowered roses

This group includes the hybrid perpetuals and hybrid teas. They are characterised by their large flowers and generally vigorous and upright growing habit. The hybrid tea is what

Figure 4.35 The large-flowered hybrid tea rose. Remove any dead, crossed or weak growth.

Figure 4.36 Hard prune large-flowered roses.

Table 4.11 Large-flowered roses (hybrid teas and perpetuals)

Amorosa
Blue Moon
Dainty Bess
First love
Flamingo
Frau Karl Druschi
Jacaranda
Just Joey
La Reine
Lorraine Lee – the exception, as it flowers in winter, prune after flowering.
Mr Lincoln
Mrs John Laing
Oklahoma
Ophelia
Pascali
Peace
Red Devil
Super Star
Susan
Whisky Mac

is usually met with in the local garden centre and produces long petalled large flowers with a high centre, a pointy shaped rose when just opened. They have become the modern idea of what comprises a 'traditional' rose. In fact, they are the result of thousands of years of breeding and are the staple of florists (see Table 4.11).

Large-flowered roses benefit from hard pruning in late winter or early spring depending on climate. Prune back the main stems to between 25 and 60 cm (depending on the vigor of the cultivar) and shorten the side growths to two to three buds. This growth will carry the next season's flower. Growth that is thinner than a pencil is unlikely to support flowers, so remove the weak and spindly as well (see Figure 4.36).

Deadheading throughout the season helps to encourage fresh bloom. In autumn, shortening the flowered stems and the removal of any crossed or dead wood will tidy

Figure 4.37 Rejuvenate large-flowered roses over a few years.

the bush for winter (see Figure 4.31). Over time, the main canes will need renewing. Cut down a third of the old stems to 10 to 15 cm and the rest staged over the next two to three years (see Figure 4.37).

Table 4.12 Cluster-flowered roses

Large shrubs – Floribunda and David Austin
Abraham Darby
Bonica
Chaucer
Graham Thomas
Gruss an Aachen
Heritage
Iceberg
Mary Rose
Pat Austin
Pink Parfait
Queen Elizabeth
Sally Holmes
Sparrieshoop
Squatters Dream
Tradescant
Winchester Cathedral
Small shrubs – carpet roses, polyantha and miniature
Ballerina
Easter Morn
Green Ice
Seafoam
The Fairy

Figure 4.38 Floribunda, David Austen and polyantha roses all respond to the same pruning regime. It can also be applied to miniatures if you can stand the fiddle. Always remove dead or crossed wood.

Figure 4.39 Late winter/spring prune for floribundas, David Austen and polyantha roses.

Remember that the harder you prune, the bigger the flowers, but also the less bush you have for that space in the landscape.

Cluster-flowered roses

Roses that bloom in clusters of buds that open at varying times are a landscape stalwart. They include the floribundas, David Austen roses, polyanthas as well as the miniature and so-called carpet roses (see Table 4.12).

In the case of the larger growing floribundas and David Austens, see Figure 4.38.

The main stems of these types of roses can be cut to about 25 to 45 cm from the graft in late winter/spring. Graduate the stems so that growth is produced at the base as well as allowing for some height. Shorten the side growths to about two to three buds and cut out any weak, crossed or dead wood (see Figure 4.39).

As with the hybrid teas, renew the rose by reducing a third of the old stems to 10 to 15 cm (see Figure 4.36).

Polyanthas, miniatures and carpet roses are the bedding plants of the rose family. Low in stature, they have proportionally smaller stems. Deadheading is always beneficial, but structural winter/spring pruning can be fiddly. It is important to remember that these roses thrive under the tender ministrations of council workers wielding brush-cutters and other mechanised pruning equipment. Those pruners who wish to can follow the same pruning technique for the floribundas and David Austens; however, these smaller roses

Figure 4.40 A very straggly neglected miniature rose.

Figure 4.41 Shear off the top two-thirds.

can be simply sheared off by about two-thirds of their original height with no ill effect. This approach is especially rewarding with miniature roses. Of course, the removal of dead, crossed and twiggy wood will be of benefit (see Figures 4.40 to 4.43). They need only about a fifth of their old wood reduced to a few centimetres in order to renew the bush on a regular basis.

Figure 4.42 Clean out the dead wood and twiggy growth.

Figure 4.43 The end result.

Old-fashioned repeat-flower roses

This group of roses includes the Bourbons, Portland roses, repeat-flowering hybrids of species roses and the old China roses (see Table 4.13). They were the first repeat flowering roses and the forerunners of the hybrid perpetuals, hybrid teas, floribundas and so forth. Until the early 1800s, roses flowered once a year in the same way as lilac, broom or viburnum does. It was the happy marriage of east and west, that is, the crossing of the ancient China rose with the roses of the west, that gave birth to the repeat flowering roses we take for granted today. If only humans could get along as well as roses.

Table 4.13 Old-fashioned repeat-flowering roses

Boule de Neige
Mme Isaac Perrier
Mme Pierre Oger
Souvenir de la Malmaison
Stanwell Perpetual

These roses need minimal pruning in winter/spring. Most carry their flowers on the side growths of healthy older wood so concentrate on thinning the shrub of crossed or weak growth rather than heavy pruning. Shorten the major stems no more than a third of their length and reduce the side shoots by half. If they are in need of rejuvenation, cut down about a fifth of the oldest flowering stems. As with all repeat flowering roses, deadheading will be beneficial (see Figure 4.44).

Figure 4.44 To prune old-fashioned repeat-flower roses, shorten the major stems no more than a third and reduce the side shoots by half. If they are in need of rejuvenation, cut down about a fifth of the oldest stems.

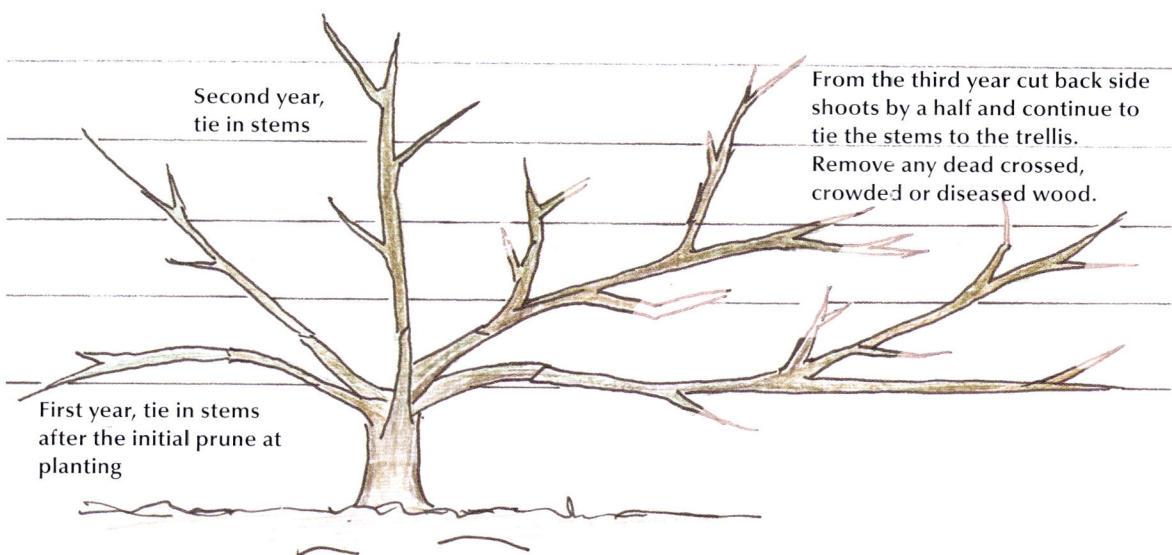

Second year, tie in stems

From the third year cut back side shoots by a half and continue to tie the stems to the trellis. Remove any dead crossed, crowded or diseased wood.

First year, tie in stems after the initial prune at planting

Figure 4.45 Pruning climbing roses.

Climbing roses

Climbing roses are ramblers that have been trained to climb or sports of existing cultivars that show an inclination to climb. The basic rule of thumb, after the initial planting prune, is to let them have their head for at least two or three years to establish a framework. *Make sure that new growth is tucked in or tied to its support as soon as it is long enough to do so.* Space the stems evenly over the surface it is to climb on with about 45 cm (the length from the tip of your fingers to your elbow) between the stems.

Do not wait until a particular season for this job; do it as you see them (see Figure 4.45). Tie the growth in as soon as it is long enough. Let them develop a framework for the first two to three years. In the fourth or fifth year shorten back the side growths by half. Renovate according to what sort of rose it is (see Tables 4.10 to 4.13). In the first year, tie in stems after the initial prune at planting. In the second year, tie in stems. From the third year, cut back side shoots by a half and continue to tie the stems to the trellis. Remove any dead, crossed, crowded or diseased wood.

Maintenance pruning at this point depends on whether the rose flowers once a year or is repeat flowering. Those that repeat through the season can have the side growths shortened by a third in winter/spring and remove any twiggy weak growth as well as dead wood. To maintain vigorous flowering wood, renew the stems by cutting back the very oldest stems to about 45 cm off the ground every two to three years depending on how congested the climber has become.

Climbers that flower once a year start with the same initial treatment. After the first few years of establishing a framework, not much is needed beyond deadheading and shortening side growths after flowering in the same fashion as for all once flowering roses (see Figures 4.31 and 4.32). Many of these climbers are extremely strong growers, however, and will need thinning rather than more considered pruning. This can be done after flowering and into the autumn. The removal of very old flowering stems every few years can be left until winter/spring. New shoots should arise from the base.

Roses on sticks – weepers and standards

Weeping and standard roses are merely your average cultivated rose grafted onto a long rootstock or interstock to give them the height above the ground. They should be pruned according to the pruning requirements for that type of cultivar (see Tables 4.10 to 4.13); however, there are some points that need consideration.

Standards and weepers will need sturdy staking for their entire life. They are by nature top-heavy and strong winds can easily decapitate them.

Roses on extremely long rootstocks do not tend to produce fresh shoots from the graft union as easily as those grafted as bush or climbing roses, and consequently their canopy does not renew as effectively.

Standard roses

As with all roses, aim to keep the centre of bushy 'ball' open to provide plenty of air circulation to prevent mildew and black spot. Most standards have at least three grafts at the apex of the stock. These form the framework branches and rarely put out new strong growth. Confine your pruning to

deadheading, shortening back side growths by no more than two-thirds, and clearing weak, twiggy and dead wood. Never prune more than about two-thirds of the growth off. The timing of this pruning should be dictated by what cultivar or style of rose you are growing and your climate (see Tables 4.10 to 4.13).

Weeping roses

The most successful weeping roses are grafted with cultivars featuring long lax growth so that they cascade downwards rather than point skyward. The ramblers and climbers are most successful and certainly vigorous (see Table 4.14). Don't bother to shorten the long growths as these will develop the side growth that in turn carries the flowers. Deadhead all of them and, depending on your cultivar, shorten the side growths to two to three buds. If the canopy is becoming overcrowded, cut back very old wood to within a few buds of the graft.

Hydrangeas

Hydrangeas, whether you call them hortensias, mop heads, lacecaps or 'Pee Wee', are favourite garden plants throughout the world. Despite their 'dying swan' act on hot days, hydrangeas are tough shrubs well worth inclusion in the landscape. Like many cultivated plants pruning is seen as essential; a way of making the plant seem dependent on human intervention. This is not so. Hydrangeas will grow and prosper with no pruning, but you will end with an enormous plant with a very small flowers dotted about. This may or may not be better than the small shrub with a few enormous blooms, as Christopher Lloyd said,

Table 4.14 Climbing roses

Climbers that flower once
Alberic Barbier
Albertine
Altissimo
Clair Matin
Constance Spry
Dorothy Perkins
Felicite Perpetue
Gloire de Dijon
Lamarque
R. leavigata
R. dupontii
Rambling Rector
Wedding Day
Repeat-flowering climbers
Altissimo
Clb Gold bunny
Clb Iceberg
Crepuscule
Dublin Bay
Golden Showers
Lady Hillingdon (Climbing)
Mermaid
Mme Alfred Carrier
Nancy Hayward
New Dawn
Pierre de Ronsard
Sparrieshoop
Twilight glow

Figure 4.46 The spent flowers can be left to overwinter on the bush. Not only do they add a different texture to the landscape, they can protect the new buds from frost.

looking more like weapons than flowers. The choice belongs to the pruner: the harder the pruning, the bigger the flowers.

There are two pruning methods for hydrangeas depending on the species you are growing.

Hydrangeas: group one

This group includes *H. macrophylla* and cultivars; the mop heads or hortensias, the most popular of hydrangeas. Also pruned in the same manner are *Hydrangea serrata*, *H. aspera* and *H. quercifolia*, known as the lacecaps and the oak leaf hydrangeas.

These species flower from buds that were formed the previous season. Like most shrubs that flower once a year, they can be cut back after flowering; however, the beauty of the faded flowers will be lost (see Figure 4.46).

Timing

In extremely cold climates, pruning is not undertaken until well into spring so that the new growth will not be cut down by frost. The flower heads can be left on the bush to shelter the new buds and subsequent growth. This has the added advantage of providing a sculptural winter presence. Those in mild climates can prune in late winter.

Technique

In young plants all that is necessary is to remove the old flower head and cut to the uppermost pair of fat buds as shown in Figures 4.47 and 4.48.

After four to five years, the bush may become overcrowded and the older growth will need to be removed. Cut down the oldest flowering stems to a strong

Figure 4.48 These buds, from last year's wood, produce the growth that will hold next season's flowers.

unflowered shoot, or if there are none, cut the stem out completely. They will be much thicker than the newer stems with pale bark. Prune out such stems at ground level. A pruning saw allows greater access, so that you do not have an awkward cluster of old stems at the base. These can make the clean removal of old stems difficult to get at in the future (see Figures 4.49 and 4.50).

A very old hydrangea can benefit from complete canopy removal if required.

If you are dealing with an ancient neglected shrub, radical renovation may be required. The entire bush can be cut to the ground, completely decapitated. This will stimulate new growth from the base. In cold climates where pruning is performed in spring, the next season's flowers will be sacrificed. In mild climates, however, where the pruning can be done in winter, flowering may be delayed. There will be enough time for the

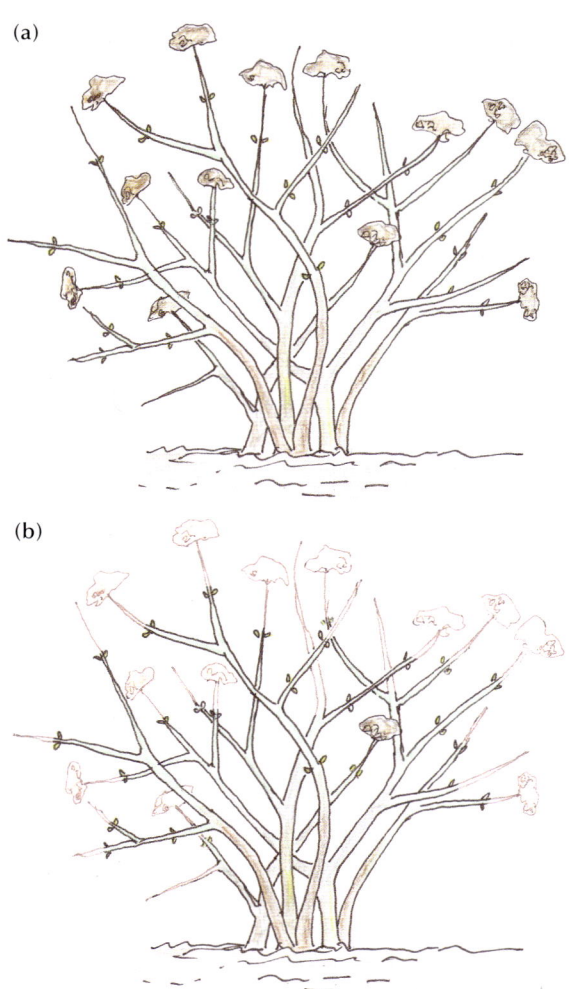

(a)

(b)

Figure 4.47 Remove the old flower heads (a and b) back to a fat pair of buds or new growth

Figure 4.49 To renovate an older shrub, remove very old stems and shorten back the younger ones by a third.

Figure 4.50 Old stems should be cut out completely to avoid this tangle of dead stems. It is easy to identify the old stems as they are pale and crusty as opposed to the darker young stems. A pruning saw can allow better access in this tight spot.

new growth to ripen and produce flower, just later in the season.

Hydrangeas: group two

Hydrangea paniculata and *H. arborescens* and their cultivars, all flower on the current season's wood; that is, the buds that are produced in spring will carry the flowers of that season. The time of pruning is the same as for other hydrangeas, dependent on climate (see Figure 4.51).

Prune back the stems depending on how large you want your shrub to be. These shrubs can be cut to the ground safely once established, but it keeps the shrub small. Alternatively, prune the stems by about a

Figure 4.51 The *H. paniculata* and *H. arborescens* species are typified by their thin stems. They flower on the wood formed on the current season and require harder pruning.

Figure 4.52 Prune the stems to the lowest set of fat buds. This keeps the shrub small.

Figure 4.53 For a taller shrub, reduce the main branches and prune back to the new growth close to the stem.

third to the lowest pair of fat buds or new growth (see Figure 4.52).

To maintain a larger shrub, prune back close to a permanent framework of branches (see Figure 4.53).

As always, the harder you prune the larger the flowers but the less shrub there is.

Climbing hydrangea

The climbing hydrangea *H. anomala* ssp. *petiolaris* can be used to cover walls, stream down banks, ascend trees or form a tangled mound as it climbs on itself. If you are fortunate enough to live in a cool to cold area, this plant is extremely vigorous once established. This hydrangea flowers on the current season's growth and can be pruned

back in spring to remove wayward branches and thin any overcrowded wood. The old flowered stems can be pruned back to within a few buds of the major branches that become deliciously brown and peeling as they age. It also responds well to complete decapitation if it has become too large for its space.

Pollarding and coppicing

These two ancient techniques really work on the same principle. Initially this pruning practice created working trees in much the same way as managed fruit trees. As a group they are both referred to as orchards. Pollarding is when a tree is repeatedly cut back to a particular point on the trunk to produce a mass of new stems (see Table 4.15

Table 4.15 Trees suitable for pollarding or coppicing
All the species listed have the ability to regrow from old wood. It should be noted that coppiced Eucalypts are more successful in climates cooler than their homeland where their growth is more reticent.

Trees for pollarding
Catalpa bignonioides Catalpa
Fraxinus spp. Ash
Gleditsia tricanthus Black Locust
Ilex spp. Holly
Lagerstroemia indica Crepe myrtle
Morus spp. Mulberry
Paulownia tomentosa Foxglove tree
Platanus spp. Plane tree
Salix spp. Willow
Tilia spp. Linden
Ulmus procera English elm
Coppicing
For foliage, ornamental stems and wood production
Betula spp. Birch
Castanea sativa Chestnut
Cornus alba Tartar dogwood
Cornus stolonifera 'Flaviramea'
Corylus avellana Hazel
E. cinerea Argyle-apple
E. perriniana Spinning gum
Eucalypt, Mallee species
Eucalypus pulverenta Silver mountain gum
Fraxinus exelsa aurea Golden Ash
Salix spp. Willow

for suitable species). The new whippy growth could be harvested for basket-making or woven into hurdles for stock control. Sometimes it provided stock fodder, or when older was used for firewood or construction timber, depending on the species.

It meant that wood could be harvested without killing the tree. Although it is still practised throughout the world for these reasons, I will be exploring its use in domestic or urban landscapes.

Plants suitable for this treatment have the ability to produce growth from old wood either from buds beneath the bark (epicormal buds), or from lignotubers as in the case of eucalypts. It is best to start on youngish trees, just a few years old, as their wounds are smaller and heal quickly. Older trees cut back this hard may not survive. In mature trees, the heartwood will be exposed and rot may set in, killing the tree.

This pruning method needs commitment. The new growth is often poorly attached to the tree; if it is allowed to mature it becomes a dangerous, unstable branch. Whether you pollard or coppice a plant for the beauty of its winter stems, juvenile foliage, to control its size or just to create a bizarre vegetable sculpture, pollards especially need regular pruning. *As so much of the plant's canopy is removed, coppices and pollards need plenty of water and nutrients to stay healthy.*

Pollarding

Pollarding is generally practised as a means of reducing the tree's canopy in confined spaces. Many suggest it would be better to choose the right-sized plant for the space at planting; others enjoy the look. Pollarding is an aesthetically acquired taste. Essentially it comprises a clear trunk and an enlarged stub or stubs from which the masses of new growth springs. The length of trunk was desirable in agricultural systems to prevent cattle eating the new growth. Pollarding is from the Middle English word 'polle', meaning 'head', a reference to the swollen branch ends. They look like something dreamed up by Dr Seuss.

Figure 4.54 Pollarding.

When your tree has reached the required height, at least 1.5 m, remove the apical bud (see 'Apical buds', page 3) of the central leader (main trunk) this will concentrate the plants resources at this point (Figure 4.54, **red bar**). Shorten back any side branches below the desired height to about 30 cm to help thicken the trunk; these can be removed completely in a year or so (Figure 4.54, **yellow bar**). If your pollard is like the Hydra, to have many heads, select the branches at the top of the trunk that spiral around the leader and have good wide branch angles. These will be the main scaffold branches (Figure 4.3). Remove any crossed branches and reduce the remaining ones to about 60 cm to 1.2 m from the trunk, depending on how you wish to train the tree (Figure 4.54, **green bar**). Prevent the main leader reasserting itself and prune out any side growths from the scaffold branches.

Once the scaffold branches are established, simply prune off the massed new growth at the end of the dormant season to a point just above the previous season's growth, about 2 cm (Figure 4.54, **green bar**). This can be done every year or every few years and slowly the 'heads' will develop.

Removing a tree's entire canopy annually will place huge demands on its economy. Ensure that it is well-watered and fed to cope, or perhaps stagger the pruning removing two-year-old growth annually.

Coppicing

Coppicing involves regularly cutting a tree or shrub to a stump or stool. This is essentially the same as pollarding but with a vastly different look Coppicing can also be used to create multi-stemmed trees. These may have showy bark (see Figure 4.55).

Figure 4.55 This silver birch was coppiced when only a few years old to accentuate the beautiful bark.

Essentially there is no trunk and therefore trees can be made to behave as a shrub. It entails cutting the central leader to within one or two buds of the ground creating a forest of slender side stems.

Coppicing was practised in English forests to open them to the light, allowing better pasture for game. Forage from the new growth of the coppiced plant and the diversity of herbage that emerged as a result of increased light conditions, encouraged deer as well as biodiversity. It is the same theory used by Indigenous Australians with their use of fire-stick farming to refresh grasses and provide pasture for kangaroos and other marsupials.

These days, coppicing is generally used to show off attractive stems or to retain juvenile foliage. However, there are still some farmers who coppice eucalypts in order to grow fence posts. The dense regrowth from the base of the plant ensures long, straight growth ideal for many carpentry needs.

This method is still used to raise selected apple rootstocks. In spring the stool is gradually covered with compost allowing the tips of the new growth to protrude. By the end of the growing season the slender regrowth will have developed roots of its own. They are then harvested and planted out ready for grafting.

When a young tree has become established at two to three years old, cut back the main stem to two to three buds from the ground. This will force a multiplicity of side growths. In the case of species with ornamental stems (see Table 4.15), this may be done every year or so to retain the juvenile bark characteristics. Those plants grown for their juvenile foliage, such as eucalypts, may need all the stems pruned back every three or four years. As soon as your plant shows signs of its middle-aged form of bark or foliage, cut back the stems to within a few centimeters of the stool. This will ensure a new era of youth and beauty.

Hedges

Hedges can be used as the skeleton of a successful landscape. They create a basic framework that can define and enhance outdoor space. They may provide a backdrop, a screen or simply a division of one space from another. Formal or informal, deciduous or evergreen, hedges may confine a fluffy flower garden, enhance sculpture or be a serenely green constant in a changing landscape.

Tools

Hedging shears are the traditionalist's first choice. They allow a degree of control that powered hedge trimmers do not. Keep them sharp (see Figure 3.13, page 36) and no other implement can give you a better finish.

Powered hedging shears/clippers are certainly quick and easy. However, it is also easier to unintentionally cut out a chunk of hedge that will take some time to grow over. Always use a gentle sweeping action to prevent the clippers 'grabbing' on to thicker branchlets. When the clippers get caught it can cause a sudden 'jump' and a possible unsightly dent in a hedge's smooth surface.

In public landscapes, bar mowers mounted on a tractor can make short work of extremely long hedges.

Whatever you use, always keep the clippers parallel to the hedge surface and clip from the bottom of the hedge up. This means that the cut material will fall cleanly to the ground; not get entangled with the longer uncut growth (see Figure 4.56).

Planting a hedge

Plant selection

Whether you are planning a formal or informal hedge, the site preparation is the same. Choose your desired species carefully

Figure 4.56 When clipping your hedge always keep the shears parallel with the hedge. Clip from the bottom up so that cut material falls cleanly to the ground.

as to height, width, climatic compatibility and the space you have available. Remember that hedges above 1.6 m high will need scaffolding in order to clip them.

If you are planning a formal hedge, make sure that you select plants that have been propagated asexually; that is, they are a named cultivar or cloned selections (see 'Propagation and landscape use', page 21). This ensures that the plants are all genetically identical and that their growth rate and habit will be uniform. Seedling-grown plants will vary ever so slightly in their leaf, flower or growth rate and will not produce the uniform seamless wall of foliage that is the aim. Sexually propagated plants are a bit like having children; they may have the same general attributes in common, but each is unique with subsequent varying strengths and weaknesses. Formal hedges are all about uniformity.

Figure 4.57 This hedge has lost its basal foliage as the upper part of the hedge overhangs the base so there is no access to light. Low light conditions have weakened the growth. Cypress is prone to this problem even when well cut.

A successful hedge should be clothed from top to bottom in foliage presenting a unified and unbroken surface. Select from species that can maintain their basal foliage to prevent an unsightly gap at ground level developing (see Figure 4.57). Such gaps will ruin the visual effect of the hedge and create wind tunnels detrimental to adjacent planting.

Formal hedges are best selected for their foliage. Shrubs with showy flowers are best for informal hedges. Flowers will interrupt the crisp line of a formal hedge, or if forced into submission, will poke out awkwardly from the foliage outline.

Hedging plants should be as tough as possible. When so many individuals are grown close together with little air circulation, pests and diseases can run rampant. Just like children at a crèche; if one catches something, they all do! For this reason I have not included species in the tables prone to fungal diseases or mite infestations. *Myrtus luma, Luma apiculata, Viburnum tinus* and *Choisya ternata* are all frequently recommended as hedging plants, but are all subject to pest and disease problems.

Site preparation

Most hedges are planned to be long-term members of the landscape, so there is really only one opportunity to get your hedge off to good start. They are a landscape structure and site preparation needs the same consideration as the foundations of a building. Mark out the site and prepare a trench the proposed length and width of the established hedge. Remove all perennial weeds, enrich the soil with compost and manure and check that the drainage and soil reflect the optimal conditions for the species you are planting. Try to create uniform soil conditions for all the members of the hedge.

Flexible irrigation pipes that can be moved out to the drip line (see page 45) of the hedge as it grows can also be planned now. Good soil preparation will ensure strong growth so that the individual plants will mesh quickly into one community – a hedge.

Always allow some access space either side of the hedge for maintenance. If you have selected a plant with extremely vigorous roots, you may also want to install a root barrier to a metre deep so all garden elements can thrive undisturbed.

Planting distances or how to space your plants

Everyone wants their hedging plants to fuse together as soon as possible. The hedge will have clear access to water and nutrients on two sides, so planting distances can be reduced from what is usual. Successful planting distances are dependent on the species used and the climate of the particular landscape. Those in areas prone to drought should allow greater distances than those in wet climates. Close spacing will reduce the vigour of plants in the long term – a desirable outcome that will reduce clipping frequency. However, adequate soil moisture must be available.

Hedges intended to reach 2 m high can be planted with 75 cm to 1.2 m between the centres of each plant. Small hedges to 1.2 m tall can be planted at 50 cm to 1 m centres, and dwarf hedges as close as 30 cm between plants. Larger distances between plants will result in a longer period before the hedge closes together, but at reduced cost of plant materials.

Initial pruning at planting

It is best to select young plants with many branches originating at or near their base. Unless the plants have been grown specifically for hedging this is unlikely to occur in advanced plants.

All hedges should be trained to be wider at the base tapering to their top. An 'A' shape is ideal. Nevertheless, practice can be different. If your region is warm and sunny, a more vertical slope can be successful. A hedge wider at the base is aesthetically more pleasing, adding an air of stability to the

Figure 4.58 The top of the hedge should always be narrower than the base so that the whole face of the hedge can receive light. (Photo taken at Heronswood, Dromana, Victoria)

landscape (see Figure 4.58). It is also essential for a well-clothed hedge. Each side of the hedge should have access to the maximum amount of light in order to ensure growth from top to bottom. In order for the foliage at the bottom to receive enough light, the top of the hedge must graduate back from the base. A batter of 5 to 10 cm for every 30 cm of height should be the minimum for areas at the 40 Parallel and beyond (degrees north or south of the equator). A more vertical incline is satisfactory between the equator and the 40th Parallel. If the hedge develops an overhang at the top, the basal growth often dies leaving a nasty stretch of dead wood, or worse, a wind tunnel underneath (see Figure 4.57). In order to induce good dense basal foliage, prune back the side growths by about at least a third to encourage the stems

Figure 4.59 Pruning a hedge at planting.

to produce plenty of branching. In all hedging plants *except the conifers,* the central leader, or main vertical stem can be cut back by a third.

If a conifer's central leader is pruned it may not grow any taller, so leave the central leader

until it has reached the desired height (see Figure 4.61).

As you will have removed so much of the plant's canopy, make sure your plant is well-watered, fed and mulched to maximise the potential new bushy growth. The fastest

Figure 4.60 Second year pruning for hedges.

Figure 4.61 Conifers should never have their central leader cut back until the plant has reached the desired height. Prune back only the side growths.

growing species can be pruned harder than the slow.

In the second year prune back the new growth by about a half at the end of the dormant season to initiate even more twiggy growth before the hedge has its first clip in mid to late spring (see Figure 4.61).

How often it needs clipping will depend on what sort of hedge you have planted and what degree of formality you require. Usually once or twice a year will suffice, with the exception of extremely fast-growing species. *Lonicera nidita* or Oz box needs maintenance up to five times a year to maintain a formal finish.

Formal hedges

The elegance of formal hedging is undeniable. Although made up of individuals, the hedge should 'read' as a single entity, presenting an unbroken and continuous surface. As always, the first consideration is plant selection, what is suitable for your site and what size you want your hedge to be.

The most formal of hedges are made up from species capable of twiggy growth on the ends of their branches with very small leaves. Large-leaved species tend to look ravaged after trimming, presenting so many cut leaves that need to heal; small leaves can hide their pain. Formal hedging plants are often unprepossessing as individuals, but acting as a group these horticultural dullards can pack a heavy landscape punch (see Table 4.16, page 95). When selecting your plant material, keep in mind how easy it will be to renovate in the future. Those species that can produce new growth from leafless wood are the simplest to manage.

Figure 4.62 This cypress hedge has grown well out over the footpath so the only alternative to the removal of the hedge is for pedestrians to duck. Cypress cannot be pruned hard as it will not regrow.

Figure 4.63 The top of a hedge can be crenulated, either formally or informally, to add another dimension to the hedge – in this case the play of light and shade. Photo taken at Ashcombe Maze, Shoreham, Victoria.

Many conifers, a classic hedging choice, do not regrow from old wood. By definition they get larger and larger with no way of rejuvenating them. If they outgrow their space encroaching on paths, driveways or garden beds there is only one traumatic solution – remove and replant (see Figure 4.62).

Other species have the ability to regrow from bare wood. These are marked with an asterisk on Table 4.16. Needless to say, these are the easiest to manage as they can recover from the procrastination, laziness or ineptitude with which they are treated. Those that do not regrow are resolutely unforgiving of human foibles.

Formal hedging styles

Formal hedges need not always be straight sided 'A' shapes. The top of the hedge should always be narrower than the base but the sides may be curved or the top crenulated (see Figures 4.63 and 4.64).

Often an archway through the hedge is desirable. This is easily achieved by leaving the requisite gap in the planting and training the top and side growths of the hedge over the gap (see Figure 4.65).

Select three or four strong stems each side of the opening and let them grow tall enough to cover the gap (see Figure 4.66). When the growth is still flexible and long enough, train them over a supportive structure and tie them in. The finished arch just needs to be clipped as the rest of the hedge. As there will be no light under

Figure 4.64 Hedges do not have to be geometric. These curves are soft yet crisp. (Photos taken at Ashcombe Maze, Shoreham, Victoria)

the arch, the once green sides of the archway will most probably die. Wooden slats or a decorative piece can be placed there to hide the gap.

Tapestry hedges are where different species or cultivars of plant are grown together to

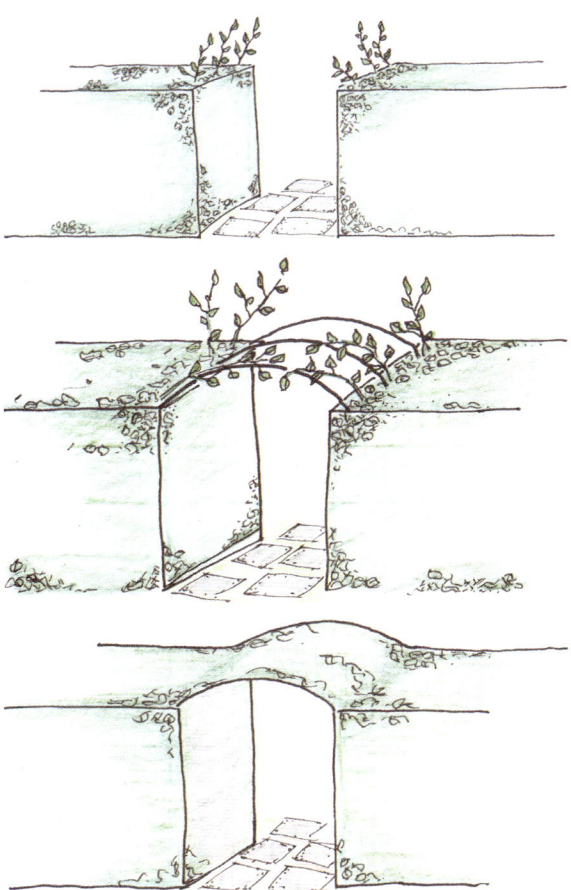

Figure 4.66 Creating an arch through a hedge.

create different foliage textures that meld together to form a tapestry of contrasting foliage. Sometimes they happen by accident when a mixed shrubbery of plants suitable for hedging become too large and are hedged to fit space requirements. The results are reliant on the species that make up the hedge. Their growth rates are of prime importance. The plants selected should have very similar growth rates otherwise one species will outgrow another making a formal outline almost impossible. The faster growing plants

Figure 4.65 An arch through a hedge can invite views from beyond or with a gate, shut out the world. (Photo taken at Ashcombe Maze, Shoreham, Victoria)

will need clipping more often than the slower growing so that the separate plants have to be clipped at varying times. Your maintenance time will be increased.

Shaping a formal hedge to a series of curves or cloudlike forms is also an effective landscape device. Simply prune to the desired shape (see 'Topiary', page 98). Again, species that grow from old wood are the easiest to manage.

Maintenance

Frequency of clipping is reliant on the level of formality desired, the species planted and the climate. The timing of clipping should be considered carefully in frosty climates as clipping after the last growth spurt if the year may promote growth only to be burnt by frost.

Those in relatively frost-free climates can achieve a crisply neat hedge for several months by hedging in late autumn. After the spring flush of growth, the outlines will have blurred so another clipping will be necessary. It is very much dependent on the species making up the hedge.

Extremely formal hedges that are expected to have a razor sharp outline need to have string lines set up to guide your clippers. Keep within these constraints and all will be well. Those less formal can be done by eye. *Always ensure that the base of the hedge is wider than the top.*

Renovating an older hedge

Always renovate your hedge, be it formal or informal, in stages. Renovation is only successful with species that produce new growth from old wood, or in informal hedges, produce new woody stems from the base. Such species are marked with an asterisk in Table 4.16. Informal hedging plants that can produce new stems or respond to decapitation are marked with an asterisk in Table 4.17 (see page 96). Many conifers do not respond so there is little to do except remove the hedge and replant.

Whether your hedge has become too large for its space or just straggly, severe renovation can be needed. This procedure will only be successful with species that can be cut to old wood and regenerate. Always renovate in stages over a few years.

Cut back only one side of the hedge at the end of the dormant season to a base of branches. Keep the hedge well-watered and manured to speed recovery growth.

Shorten back the new growth after the first burst of growth to promote bushy growth. Clip the developing hedge as required.

When the renovated side has recovered full foliage coverage renovation of the other side can be undertaken. This may not occur for two to three years.

Informal hedges

The difference between screen planting and an informal hedge is a moot one. Suffice it to say, a hedge is usually made up of the one species. They will act as one entity intertwining their branches so that one shrub is indistinguishable from another. Again the golden rule is to have the base of the hedge wider than the top. A broad-based hedge

Figure 4.67 Renovating a hedge.

looks stable and grounding, one with a distinct overhang at the top becomes a series of bare stems with the beauty and foliage out of visual reach; in fact, an avenue or border of small trees. If there is no light at the base there will be no foliage.

As for formal hedges, prune at planting to create bushy growth (see Figures 4.59 and 4.60). In their second year, let them develop and only in the second dormant season shorten back wayward branches and shorten the stems to encourage plenty of side shoots.

After this time prune your hedge as if it was a single informal shrub and according to its type (see 'Ornamental shrubs', page 51). The timing of clipping or shortening back branches will be dependent on when your shrub flowers or fruits, if that is the reason for

its selection. Hedging plants that may become pest plants can be sheared before their seeds or fruits ripen and disperse. Otherwise neaten up as you see fit. Autumn is a good time in areas without frost; after the spring flush of growth in cold climates.

HEDGES

- Plants that regrow from old wood are the easiest to manage
- Always have the base of the hedge wider at the base than the top
- Choose cloned (asexually propagated) plants; they will grow at a uniform rate
- Keep clippers/shears parallel to the surface of the hedge
- Clip/shear from the bottom up
- Renovate only one side of a hedge at once

Table 4.16 Plants for formal hedging

Large – to 2 m plus *can be pruned into old wood
Evergreens
Also suitable for informal hedging
Acmena smithii Lilypilly
Buddleja spp. Butterfly bush
Callistemon spp. Bottlebrush
Escallonia spp. Escallonia
Griselinia littoralis Griselinia
Ilex spp. Holly
Ligustrum spp. Privet
Olea europa Olive
Pittosporum spp. Pittosporum
Prunus lusitanica Portugese Laurel
Syzgium spp. Lilypilly
Taxus baccata Yew
Berberis spp. Berberis
C. torulosa Bhutan cypress
Cupressus macrcarpa Cypress
Garrya elliptica Garrya
Leptospermum leavigatum Coastal teatree
Photinia robusta Photinia
Deciduous
Carpinus betulus Hornbeam
Crataegus oxycantha Hawthorn
Pyracantha spp. Fire thorn
Continus coggygia Smoke bush
Medium – 1 to 2 m
Evergreens
Abelia grandiflora Abelia
Buxus sempervirens English box
Cotoneaster spp Cotoneaster
Grevillea rosmarinifolia Rosemary grevillea
Artemesia arborescans Tree wormwood
Coleonema album Diosma
Hebe salicifolia Willow Leaf Hebe
Lonicera nidita Oz box
Myoporum insulare Boobialla
Myrtus communis Myrtle (can be prone to mite infestation when stressed)
Rhododendron indica Azalea

Rosmarinus officinalis Rosemary
Westringia fruticosa Coastal rosemary
Dwarf hedges – 1 m and below *can be pruned into old wood
Evergreen
Buxus microphylla Small leaf box
Buxus sempervirens 'Suffruticosa' Elite box
Coleonema pulchrum Dwarf diosma
Hebe diosmifolia Box leaf hebe
Myrtus 'Tarentina' Dwarf myrtle
Santolina chamaecyparissus Lavender cotton
Helichrysum italicum Curry plant
Informal hedges – 2 m and above (*See also plants under formal listing*) *can be pruned into old wood
Evergreen
Camellia spp. Camelia
Citrus spp. Citrus fruits
Feijoa sellowiana Pineapple guava
Fortunella spp Kumquat
Hibiscus rosa sinensis Hibiscus
Metrosideros tomentosa New Zealand Christmas bush
Nerium oleander Oleander
Persea americana Avocado
Psidium littorale Strawberry guava
Solanum rantonettii Potato bush
Tecomaria capensis Tecoma
Ceonothus spp. Californian lilac
Ceratonia siiqua Carob
Laurus nobilis Bay tree
Plectranthus eklonii Shrubby plectranthus
Deciduous
Lagerstroemia indica Crepe myrtle
Sambucus nigra Elderberry
Syringa vulgaris Lilac
Chaenomeles japonica Japonica
Corylus avellana Hazel
Punica granatum Pomegranate
Dwarf 1 m and under
Caryopteris clandonensis Caryopteris
Daphne odora Sweet Daphne
Felicia ammeloides Agathea

Table 4.17 Informal hedges *can be pruned into old wood

Large – 1 m to 2 m
Evergreen
Alyogne huegelii Native hibiscus
Carrissa grandiflora Natal plum
Cistus spp. Rock rose
*Citrus spp. Citrus fruits on dwarfing rootstock
Correa alba White correa
Eriostemon myoporoides (syn *Philotheca*) Wax flower
Heliotropum arborescens Cherry pie
Leonotis leonorus Lion's ear
Murraya paniculata Orange jessamine
Phlomis fruticosa Phlomis
Thryptomene Thryptomene
Ugni mollinae Chilean guava
Deciduous   *can be pruned into old wood
*Aloisya triphylla Lemon verbena
*Hydrangea spp. Hydrangea
*Rosa rugosa Rugosa rose
L.scoparium Manuka
Lavandula spp. Lavender
Ribes spp. Flowering burrant, red currant, black currant, gooseberry (can produce new growth from the base)
Spirea cantoniensis May bush (can produce new growth from the base)
Symphoricarpus spp. Coral and snow berry (can produce new growth from the base)

Pleaching

Pleaching is a form of living architecture. It is an ancient art traced back to the Romans and involves plaiting or weaving living branches together to form a structure. In the process of training or over time, these branches fuse or are grafted together. This means that the cambium layers (see pages 1 and 2) of individual plants unite forming a self-supporting structure. The most advanced pleaching techniques involve exposing the cambium layers of the woven stems; that is, removing the outer bark so that the rich green cambium of both stems

Figure 4.68 The cambium layer of the tree can be selectively joined to create amazing tree shapes. This example by Pooktre may take a bit of practice. It does show the possibilities. (Photo courtesy www.pooktre.com)

can meet and unite. These structures can then take the form of arbors, arches and tunnels, or can form more ambitious configurations (see Figure 4.68).

Figure 4.69 A high-rise hedge. A pleached arbor in the Jardin du Palais Royal in Paris with its magically dappled light.

A type of hedge on stilts is what is usually met with; a kind of high-rise hedge (see Figure 4.69).

The plants are grown to form a thin wall of foliage used for screening, windbreaks, to define a space or to provide shade in a confined space. It is closely allied to the art of espalier (see page 131) and some fruit trees can be grown in this manner. Select trees that have flexible branches with a strong central leader or main trunk. Planting young stock such as a whip (see Figure 5.1, page 125) ensures easy training. The ability to spontaneously graft together (inosculate) is an advantage (see Table 4.18).

To create a pleached structure, select trees of the same age and cultivar so that growth can be as uniform as possible. If you are after foliage coverage, plant your trees running in a north to south line so that both sides of the form obtain maximum sun. If you are after fruit production, planting in an east–west line will prolong harvest time. Many sites may not give you the choice, especially if the structure is intended to screen, so growth may be uneven with varying light levels. This will only mean that pruning will have to be adapted to suit.

A solid framework should be constructed using poles together with tensioned wires or wooden dowels/bamboo. There should be a post for each tree except for the first and the last post. Depending on the size of the tree or shrub to be pleached, planting distances can vary from 1 m to 2 m to 2.4 m between plants. Allow approximately 3 m between rows if a tunnel or alley is the object. Tie the trunks to the posts and fasten the branches to your framework. The branches below the desired height of bare trunk can be shortened back to 30 cm to minimise the loss of leaf coverage that could hamper establishment. They will encourage a strong trunk, and can be cut out completely after a few years.

As the trees grow above the base of the wires or bamboo structure, cut back the apical bud to encourage branching. The subsequent shoots can be tied to the support until the branches interlace. Cut out completely any growth that is growing at an angle away from the desired plane. Cut back strong vertical growth and remove any branches below the designated screen (see Figures 4.70, 4.71 and 4.72)

When establishing a pleached structure, prune in winter to promote growth and train through the spring and summer. Once it has

Table 4.18 Plants for pleaching; *can be pruned into old wood

*Acmena smithii Lilypilly
*Carpinus betulus Hornbeam
*Castanea sativa Chestnut
*Escallonia spp. Escallonia
*Feijoa sellowiana Pineapple guava
*Hibiscus rosa sinensis Hibiscus
*Malus spp. Apples
*Morus spp. Mulberry
*Olea europa Olive
*Platanus spp. Plane
*Psidium littorale Strawberry guava
*Pyrus spp. Pear
*Sambucus nigra Elderberry
*Sysgium spp. Lilypilly
Laburnum 'Vossi'
Laurus nobilis Bay tree
Punica granatum Pomegranate
Tilia Linden
Wisteria spp.

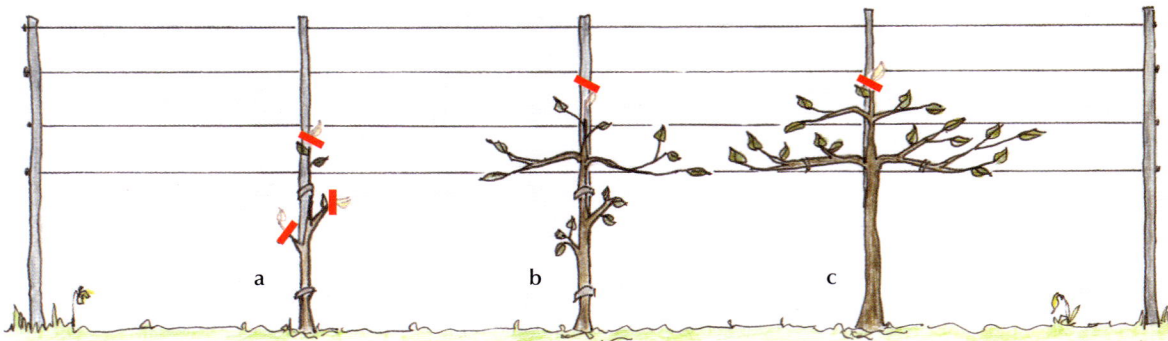

Figure 4.70 Construct your frame and tie the young tree to the posts. Shorten back the side shoots below the desired height to aid establishment (a). Cut back the uppermost (apical) bud to promote bushy growth. Train the branches to the frame (b). Continue to nip out the uppermost bud in winter and train the branches in summer. Small branches below the wire can be removed completely (c).

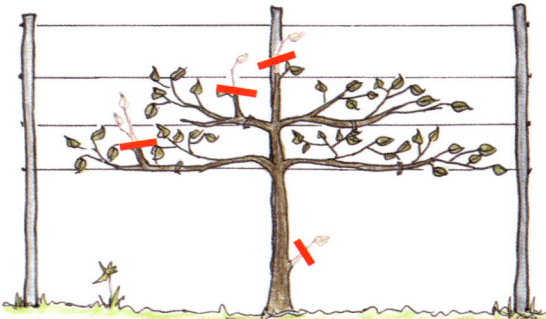

Figure 4.71 Shorten back any vertical growth and remove shoots below the major canopy.

reached fullness, pruning and clipping in summer will reduce growth and hence maintenance.

Maintain your living sculpture by simply hedging as required. To renovate an old overcrowded structure, thin older stems at the end of the dormant season.

Topiary

Topiary is the art of shaping a naturally bushy plant into a dense and solid shape. It may be a ball on a stick (a standard), a simple

Figure 4.72 Remove any growth that deviates from the two-dimensional plane of branches you are trying to create.

geometric shape such as a cone, ball or cube, or it could be something more fanciful like a

Figure 4.73 Topiary can be used to dramatise or accent any landscape. This example uses hornbeam.

hippopotamus or a bird. Whatever the design, topiary requires dedication and perseverance to achieve a solid form that can accent, anchor or dramatise a landscape (see Figure 4.73).

There is a cheat's way around the long training. An existing shrub can be 'carved' into, and then trained to become a dense shape, but this strategy will only work with plant material that sprouts easily from old wood.

Simple shapes work best so that they do not disappear between maintenance trimmings. Also, choose an open position so that growth can be as even as possible around the whole shape. Good access to sunlight for all surfaces of the shape is vital.

Plant selection

Evergreens are the usual subjects for topiary. Slow-growing, fine-textured (or leaved) plants will give the greatest scope as plants with

larger, coarser leaves cannot present the continuous leaf coverage that shows off the desired shape. The smaller the leaf, the finer and more detailed the design can be. Fast-growing plants will need much more training time if the trainer is to keep up with the growth rate, and maintenance will perforce be more frequent. Plants that sprout easily from old wood are ideal as any mishaps can be rectified – albeit in time (see Table 4.19).

Table 4.19 Plants suitable for detailed topiary
Choose cultivars with small leaves for the best results – they can also be grown as standards.

Large plants to 2 m plus (*plants that can resprout from old wood)
Acmena smithii Lilypilly
Buddleja spp. Butterfly bush
Escallonia spp. Escallonia
Ilex spp. Holly
Ligustrum spp. Privet
Pittosporum spp. Pittosporum
Syzgium spp. Lilypilly
Taxus baccata Yew
Berberis spp. Berberis
C. torulosa Bhutan cypress
Cupressus macrcarpa Cypress
Leptospermum leavigatum Coastal teatree
Photinia robusta Photinia
Quercus ilex Holm oak
Deciduous
*Carpinus betulus Hornbeam
*Pyracantha spp. Fire thorn
Crataegus oxycantha Hawthcrn
Medium plants 1 to 2 m
*Abelia grandiflora Abelia
*Buxus sempervirens English box
*Cotoneaster spp. Cotoneaster
*Grevillea rosmarinifolia Rosemary grevillea
Coleonema album Diosma
Lonicera nidita Oz box
Myoporum insulare Boobialla
Myrtus communis Myrtle
Rosmarinus officinalis Rosemary
Westringia fruticosa Coastal rosemary

Plants with showy flowers are usually trained as standards rather than the more detailed topiary shapes.

Ensure that whatever you choose is suited to the site. It is heartbreaking to spend hours of effort only to see your masterpiece destroyed by frost, sunburn or die back through drought.

Training topiary shapes

If you are wishing to train a plant into a standard, see below. The major difference between a topiary shape and a standard is that the former has a dense foliage base at ground level; a standard is topiary on a trunk.

Start with a young plant or struck cutting. With a clear idea of the shape required, clip off the ends of all the plant's stems into the shape desired, but much smaller. This will promote bushy growth and it is vital that topiary be dense. A dense framework will strengthen the ultimate structure. Continue this regimen at regular intervals so that new growth does not get longer than 5 to 8 cm long before its tips are pinched out/clipped.

If a plant with large leaves such as *Laurus nobilis* or bay tree is being used, trim back the individual stems rather than clipping the leaves, otherwise it will look ravaged.

If you are using a wire frame to guide the ultimate shape, tie stems to the frame with biodegradable string and don't neglect to constantly pinch out the tips of any stems that will form the 'body' of the shape. This may seem unnecessarily slow but it will build a solid structure to support the ultimate shape (see Figure 4.74).

Figure 4.74 If you are using a wire frame to guide your shape, tie appropriate branches to the frame with bio-degradable string and remember to pinch out the tips of *all* growth more than 5 to 8 cm long.

Standards – balls on sticks

Many standards can be bought readymade. They usually consist of a bushy or weeping plant grafted onto a sturdy rootstock that forms the trunk as with weeping or standard roses (see page 78). Highly successful and more natural-looking standards can be grown, however, when the one plant forms both the trunk and the 'ball' atop it.

Plant selection

Choose a plant with a strong central stem or leader. It may be an unbranched cutting or seedling or a more advanced shrub with the essential strong central stem that will form the standard's trunk. Many plants with showy flowers can be grown as standards as the 'ball' on top of the long trunk is a simple shape that shows off the flowers and then can be clipped back into shape (see Table 4.20). Information about the plant selection for topiary shapes is also relevant.

Table 4.20 Plants suitable to be made into standards
All those listed under topiary are also suitable.

Large plants to 2 m plus (*plants that can resprout from old wood)
Callistemon spp. Bottlebrush
Olea europa Olive
Prunus lusitanica Portugese laurel
Garrya elliptica Garrya
Hebe salicifolia Willow leaf hebe
Laurus nobilis Bay tree
Medium plants 1 to 2 m
Abutilon Chinese lantern
Aloisya triphylla Lemon-scented verbena
Chrysanthemum frutescens Daisy
Fuchsia x hybrida Fuschia
Hebe spp. Hebe
Heliotropium arborescens Heliotrope/Cherry pie
Laurus nobilis Bay tree
Lavandula spp. Lavender

Training a standard

Growing a successful standard is all about maximising the power of the apical/top bud (see page 3). This bud at the top of the central stem manufactures the future trunk and must be kept growing vigorously. Straight after planting, insert a stake against the existing central stem and use soft ties to fasten them together every few centimetres. This will ensure that the stem grows as vertically as possible which will maximise the growth of the apical bud. Vertical wood growth is always the most vigorous, and suppresses the growth from buds below it. Strong staking will also ensure a more symmetrical outcome and support the as yet immature future trunk.

Remove or rub off any growth near ground level, and if you have an established plant, lop off the side branches to within two side

Figure 4.75 A successfully trained standard. Note the topiary hedge in the background.

growths of the apical bud. If the plant is very immature and needs all the leaf coverage it can get to establish itself, pinch back the side growths to one leaf for vigorous plants or to 2 to 5 cm of growth for those that seem to be struggling. Whatever the vigor of your plant, shoots will continue to grow from the trunk. Rub them off when they are small (see page 31) to ensure they do not regrow. Persistent side branches can be removed effectively by disbudding (see page 43).

Keep removing or pinching back side growths to within two side growths from the apical bud until the trunk has reached the desired length. Any side growths that you have kept to improve the vigour of the trunk and whole

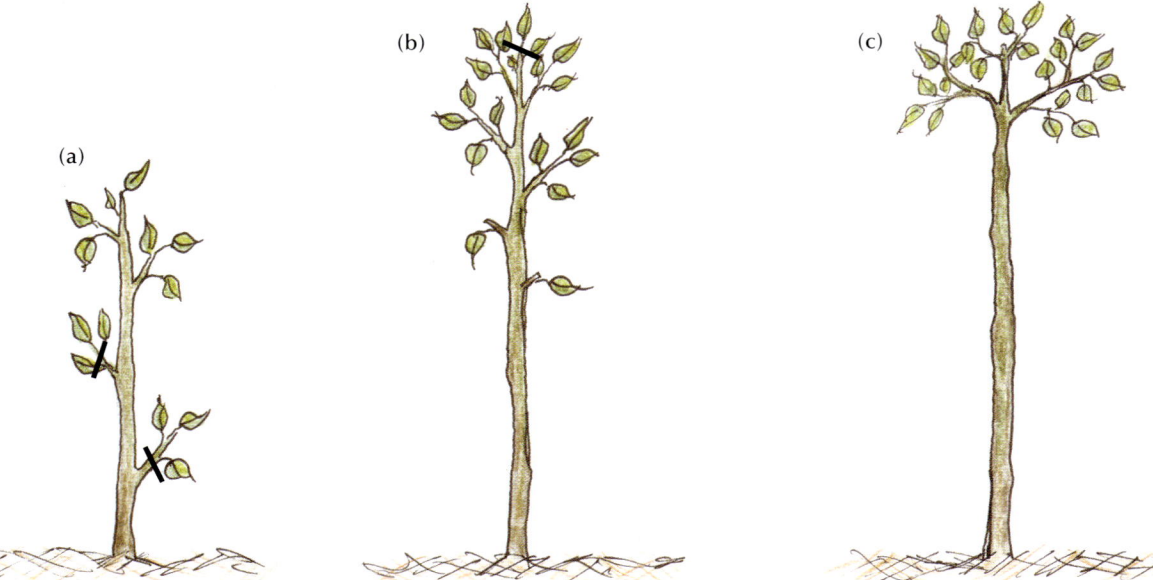

Figure 4.76 Training a standard. Select a plant with a strong central stem. Cut off side growth low on the plant and cut to one or two leaves side growth further up (a). Continue this process until the trunk has reached the desired height with two strong side growths below the apical bud. Cut out the apical bud (b). Remove any remaining side growths and pinch out the tips of any new growth 5 to 8 cm long (c).

plant can be pruned out before removing the apical bud. Now is the time to encourage the bushy growth that will form the 'ball' on top of the trunk (see Figure 4.76).

Follow the same technique as for the topiary shapes by constantly pinching out the tips of new growth. Make sure that this growth does not get longer than 5 to 6 cm, always keeping in mind the shape you wish to achieve. The end result should look something like Figure 4.75.

Renovating older trees and shrubs

Almost every garden has old and neglected trees and shrubs whether they have been under your care or not. Some have simply become too big for their space; they may have ceased to be fruitful, or failed to flower as profusely as they should.

First assess whether the plant is worth saving or not. Is it diseased with many rotting limbs or a bad viral infection (long, random yellowish streaks on the leaves; take these to be diagnosed by a horticulturalist or nurseryman)? Is it as completely inappropriate as a forest giant planted in a tiny inner city courtyard? Is it a short-lived species that has come to the end of its aesthetic life, like many quick-growing species such as *Acacia*? Has the ecosystem of the garden changed (e.g. trees grown up and shading a sun-loving species)? If so, the best solution is to remove or relocate it and replant. Remember that nothing can beat good plant selection.

Do some reading or check with your local nursery for plants that thrive in your garden conditions. Also check that they are the right height and width for your needs and will not grow into an expensive problem by damaging house foundations.

Once you have made the decision that it is worth saving, spend some time just looking at it. Whatever the plant, start by removing any dead or diseased wood, this clears the way for the next step. Consider what you want from the plant. Is it more flowers, fruit, a stately canopy or space to move around or under it? Is it a shrub that could be turned into a small tree? Can you discern a pattern in the branch structure?

Don't make large cuts all at once. It is best to make many smaller cuts than one big mistake, and don't expect to renovate your plant in one session. It is better to space your pruning over a year or two. Removing too much foliage at once will deprive your plant of its food-producing source (see page 5, 'How plants make their own food').

Summer pruning is best to restrict growth; however, if you wish to invigorate your tree or shrub, prune in winter (see 'When to prune', page 26). Usually a mix of summer and winter pruning works best. Trees have four main types of aboveground woody parts: the trunk, scaffold branches, secondary branches and lateral branches that hold the flowering/fruiting wood, or in the case of conifers, the majority of the foliage.

Renovating ornamental trees and shrubs

Many ornamental trees can be safely left to their own devices and may never need anything more than the removal of dead or damaged wood or crossed branches. However, when there are serious structural defects to correct such as competing leaders pruning can provide the solution. It is also possible to make more space under the tree, to reduce the tree's height or width, or to thin the canopy to allow better air circulation or allow more light into the garden. Hacking away at the periphery of the tree's canopy will only produce a hedged effect which will destroy the tree's beauty. This hedging technique on trees only looks acceptable when the tree is acting as a landscape group like any other hedge (see also 'Pleaching', page 94).

Before removing what appear to be excess branches, consider what sort of tree you are dealing with. Is it a shade-loving tree? If you remove too much material the exposed bark may become sunburnt. Is it a conifer? Most conifers do not grow from old wood, or wood that carries no leaf or green parts. Once the green parts are removed the branch will never regrow, so pruning of conifers should be judicious.

There are few basic shapes for a tree's canopy. There are spire-shaped trees with a central leader or main stem, a rounded crown, or weeping trees and shrubs such as weeping roses, *Betula pendula*, silver birch or weeping willow *Salix babylonica* (see Figure 4.77).

Trees and shrubs with a central leader

Trees with a strong central leader or main trunk can need the removal of rival leaders.

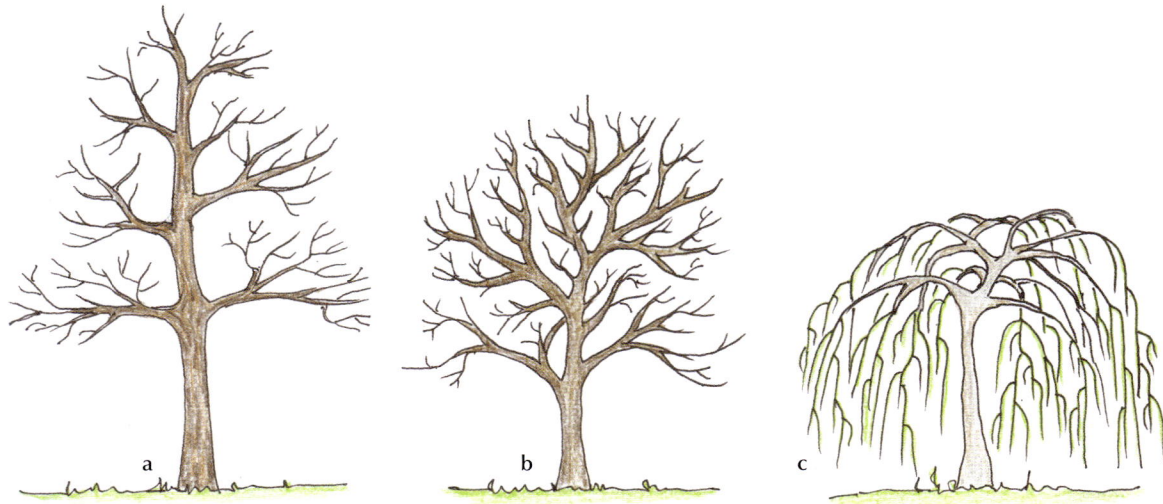

Figure 4.77 Three basic tree and shrub shapes. The symmetrical spire (a). The rounded or randomly branched (b), and the weeping tree (c).

A leader is really just an extension of the trunk, rising like a central column from the ground to the tip of the tree. Do not attempt to reduce the height of such trees. Their beauty lies in the regular assemblage of branches radiating from the central leader. Once the leader is lost the tree loses its symmetry. However, often as in life, there is a wayward branch that challenges the central leader. It should be dealt with immediately. Remove it from its point of origin at the central leader (see Figure 4.78)

Spire-shaped trees will, with age, lose the top of the central leader. This is merely a function of the age of the tree.

Many conifers conform to this general shape, and being evergreen, it can be difficult to see the branch structure. Part the foliage to have a look inside the plant. Not only will you be able to discern how the branches are placed, you will be rewarded with the manly perfumes so characteristic of conifers.

Remedial pruning beyond removing competing leaders is usually not that successful with most conifers in the long term. If they are too wide or too tall, removal of the whole plant may be the best option.

Reducing height, width or making more space underneath your tree

Reducing height

Trees and shrubs with rounded crowns or less-ordered branch structures are more democratic in their branch hierarchy and are therefore easier to modify. Spire-shaped trees can be made less dense or narrower, or to have more room underneath the canopy; however, their height cannot be altered without sacrificing the symmetry of the tree. Reducing height and width of other trees and shrubs is relatively easy. I am using a crown-shaped to random asymmetrically shaped tree to illustrate in Figure 4.79.

Figure 4.78 Remove competing leaders so that the symmetry of the plant is not harmed. Remove wayward branches from their point of origin.

Figure 4.79 Reduce the height and thin the canopy of trees and shrubs by pruning to a lower crotch or branch junction. This sort of pruning can reduce the shade produced as well as minimise wind resistance.

To reduce the height of a tree, it is necessary to reduce the height of the major structural branches by cutting them down to a secondary branch. Do not attempt this with spire-shaped trees. Always prune to an outward-facing branch. This technique is also called drop crotching. It simply means that the major branch is brought down to an appropriate crotch, or junction, with a secondary branch.

Reducing tree and shrub width

A wide-spreading tree or shrub is certainly beautiful, but few of us have the luxury of unlimited space. As always, look at your tree first. Remove any drooping low-hanging limbs. See if you can discern where the branches tilt upwards and remove the growth that is growing sideways, leaving the growth that is growing in a more upward direction. Always remove branch parts from their point of origin so the plant can heal and the limb will look like one continuous branch. This must be done to the entire canopy if the tree is to retain its balance. Wide upper branches will shade the lower limbs that have been pruned to grow upwards. This will reduce the lower limb's growth rate or even kill them. The periphery of the canopy must be graded to allow light to reach its entire surface (see Figure 4.80).

Create more space underneath your tree and making shrubs into small trees

As trees mature, lower limbs can be removed to allow the gardener space to cultivate under the tree or to provide a space for a garden seat. Removing lower branches can also open up views behind the tree, allowing glimpses of

Figure 4.80 Reduce the width of your tree or shrub by pruning to an upward growing branch.

a tempting landscape beyond. Remove some of the lower branches of your tree to create the desired effect. Do not remove too many as this will reduce the leafy canopy too dramatically and therefore reduce growth (see Figure 4.81).

Plants that can't decide whether they are a tree or a shrub can be forced to commit to a more treeish form. Many shrubs that have grown too large for their space can become heavy and oppressive in the garden. The best solution (provided it is not planted for screening purposes) is to convert it into a multi-stemmed small tree, thus freeing the ground space and creating a higher leafy canopy, in favour of an undistinguished blob. If the plant was intended to screen ugly views choose plants

Figure 4.81 Make more room under your tree by removing lower branches.

that respond well to decapitation, or see above, 'Reducing height and width'.

Pruning weeping branches

Whether it is a weeping rose, a weeping ornamental cherry or a huge weeping willow, the pruning theory is the same, just a change in scale. Often a weeping tree precludes anyone from passing underneath. It may hang itself over a fence next to a footpath, or the overall outline of the plant might become too solid. An unpruned weeping plant may no longer drape its foliage gracefully, but become a dense and impenetrable wall much like a shrub that has grown too heavy and dominating for its space. In the case of children's gardens, the latter is a decided advantage – there is no better spot for a cubby. In order to remain elegantly wistful if not mournful, however, a little thinning of the branches is often required.

Figure 4.82 Turning an overly large shrub into a tree can change the feeling of your garden and open up new horticultural horizons. Remove lower branches in much the same manner as lifting the canopy of a tree.

Figure 4.83 Pruning weeping branches without destroying the gracefulness of the tree. Never hack the bottom of the branches off evenly, you will end up with what appears to be a suspended hedge.

Never just trim off the ends of the branches evenly. This will produce a heavy fringe of foliage, much like a suspended hedge, and destroy the light elegance that is so admired in such species. Pruning lower branches is an obvious place to start (see Figures 4.81 and 4.82); however, the soft 'weepers' also need to be shortened. Look carefully at the branch and cut to an upper- and outward-facing branch. The growth on this branch can also be shortened to an outward-facing branchlet. Continue this way and your weeping plant will look naturally light and graceful. This method can be applied to any weeping plant from roses to giant willows (see Figure 4.83).

Climbing plants – ornamental and edible

Climbing plants hold a special niche in the ecology of any landscape. They are uniquely adapted to ascend to great heights without the initial woody tissue of trees and shrubs.

They may be annuals such as garden peas and beans. These plants need training rather than pruning as they only require a trellis/framework and someone to thread them to it. Climbing roses are covered under 'Roses', page 67.

Long-lived climbers will need pruning either to make them more productive, as with grapes and wisteria, or just to keep them within bounds (see Figure 4.84).

Supports

Whatever you choose to grow your climbing plant on, it must be sturdy. When fully grown, climbers can carry a significant amount of water on their leaves. This is extremely heavy and it is the support that keeps the climber up, not the plant itself. They also can act like a sail; in windy conditions they are vulnerable. So if you are dealing with a long-lived climber ensure that it has well-anchored, adequate support.

RENOVATING OLDER TREES AND SHRUBS

- Decide whether your plant is worth saving or should it be removed
- Always have a good look at your plant to discern its branch structure before pruning
- Don't make large cuts all at once. It is best to make many smaller cuts than one big mistake
- Drastic renovations should be spaced over time. Removing too much foliage at once will deprive your plant of its food-producing source (see page 5, 'How plants make their own food')
- Always prune to a branch or growth point that is pointing in the direction you want the plant to grow
- Don't expect large crops of fruit or flowers after a major renovation

Figure 4.85 Many climbers are sun lovers and race to the top of their support to gain the best access to light. Pruning at planting could have alleviated this rather ugly result.

How climbers grow

Most climbers crave the sun and are therefore keen to grow above all competition to become an elevated shrub usually out of human sight atop a forest giant. This strategy leads to an ugly mass of bare woody stems topped by a dense tangle of foliage (see Figure 4.85). This behavior is most apparent in flowering climbers.

They should therefore be planted where they get maximum sun and pruned to stimulate new growth low on the plant. Shade-tolerant climbers are more forgiving, with the ability to retain foliage over their entire surface (see Table 4.21).

Figure 4.84 These climbers are well out of control and pose a threat to their support – in this case an entire house. Jasmine and ivy are a powerful mix.

Table 4.21 Climbers tolerant of shade

Akebia quinata Chocolate vine
Ficus pumila Climbing fig
Hedera helix Ivy
Muelenbeckia complexa Wire vine

Figure 4.86 Three adaptations that allow climbers to ascend. Ivy with aerial roots cling to a solid surface (1). Tendrils that latch on to any likely support (2), and twiners that wind their way around a framework (3). Roses and bougainvillea use hooked thorns to advance.

There are approximately four ways that climbing plants climb (see Figure 4.86).

Plants such as peas, passionfruit and grapes climb using tendrils. Tendrils are modified leaves or stems that reach out to the support, and coil around it. Many are twiners, spiralling around their support such as kiwifruit or jasmine. Roses and bougainvillea use their hooked thorns to scramble their way to the top, while ivy, uses aerial roots to cling to a solid surface.

Self-clinging climbers

Climbers such as *Parthenocissus* spp. (Virginia creeper), *Ficus pumila* (climbing fig)and *Hedera* spp. (ivy) all have evolved plant parts that allow them to cling to solid surfaces. Therefore, they need a wall or solid fence to climb on; trellis or wire will not support them. They are vigorous growers that can become shrubby if left unpruned. As their stems are attached to their support they are best hedged hard back to the supporting surface.

In order to enjoy the autumn colour of Virginia creeper/Boston ivy this can be done after the leaves fall. However, they all benefit from a late spring trim, or for that matter, any time they become too heavy or rampant. They are extremely difficult to kill, but easy to manage when kept in bounds.

Contrary to popular belief, these 'suction cups' or aerial roots do not damage walls (unless they are already crumbling) but when removed the remains of their adhesive roots/pads can make the surface unsightly. Damage by climbers is generally a result of the active growing stems threading themselves into small spaces and then expanding as the plant grows; a maintenance issue.

Climbers as hedges

Climbers can also be utilised as hedges. Given a sturdy support such as heavy-gauge wire fencing, climbers can be trained to the support and clipped as necessary. Choose shade-tolerant vines so that the support is clothed from top to bottom (see Figure 4.87).

Figure 4.87 Wire vine or *Muelenbeckia complexa* makes a dense impenetrable hedge when grown on sturdy wire fencing. (Photo taken at Heronswood, Dromana, Victoria)

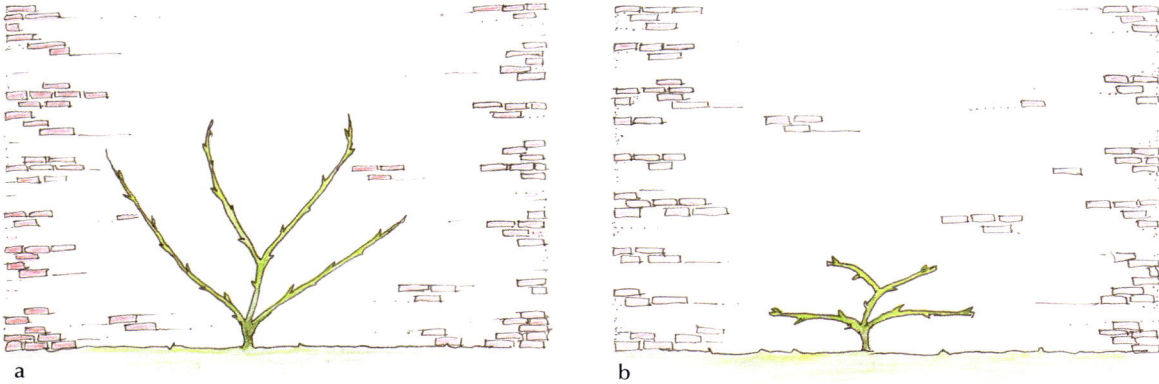

Figure 4.88 The vine has been trained to the wall without pruning (a). It is likely to be devoid of foliage at the base in a few season's time. A pruned vine with the stems trained as horizontally as possible and well-spaced is likely to retain foliage over its entire surface (b).

Thread the vines through the support. Once it has extended over the surface required, the climber can be clipped as a hedge. The same rules for the shape of hedges apply (see page 85, 'Hedges'). *Always keep the hedge wider at the base than the top.*

Pruning at planting
Fan training
If you need your climber to cover a support from top to bottom, prune back the long stems to buds facing in the direction you want growth to go. The subsequent side growths will provide a base for dense foliage cover at the foot of the plant. Space the stems evenly over the surface, bending the stems at the extreme right and left of the plant as much to the horizontal as possible (see Figure 4.88).

Climbers on one trunk
Some climbers are grown to form a trunk before the leafy canopy is allowed to develop. Grape vines are the obvious example, but wisteria and other flowering climbers can also

be trained this way to form a mini tree or standard supported by a strong stake (see Figure 4.89).

After planting, select the strongest, most upright stem and train it to its support. At the end of the dormant season prune out all side growths but leave the growth at the top to continue growing until it has reached the desired height.

Once the required length of the trunk has been attained it is time to induce branching. Prune back the top branches to stimulate side growth that will become the major scaffold branches of your plant. If you are training a plant to a two-dimensional trellis (e.g. grapes), select two branches, if you want to train the plant over a flat surface like the top of a pergola select up to four stems depending on the horizontal space available to cover. When training a climber flat, space the branches evenly as if fan training.

Only once the trunk and scaffold branches are in place can pruning begin for flowers

Figure 4.89 Training a climber to form a canopy over one main stem.

and fruit. This will depend on what sort of growth the climber flowers on.

Climbers that flower on new season's growth

This group is easy to prune and renovate, especially the long-lived species. They can be left alone to fill their space. As they flower and fruit on growth produced in the current growing season, however, they will eventually flower only at a distance from the centre of the plant which will become tangled and unproductive (see Table 4.22).

Many plants, especially clematis in this group, benefit from being cut back to within 40 cm of the ground in the dormant season (see Figure 4.90).

Other plants such as *Passiflora edulis* (black passionfruit) would not tolerate this treatment and many pruners are unwilling to be so severe. The penalty is a more time-consuming and considered approach.

In spring, cut back the new growth to about 30 cm, remove some completely to allow for good air circulation, and of course prune out

dead or diseased wood. Once the vine is established, cut back hard one major stem

Figure 4.90 Clematis can be cut back extremely hard and recover with renewed vigor. (Photo taken at 'Burnside', Ascot, Victoria. See www.lambley.com.au)

Table 4.22 Climbers that flower on new growth, prune in winter/spring

Bouganvillea spp. Bouganvillea
Clematis aristata Old man's beard
Clematis tangutica and cultivars
Clematis ternifolia and cultivars
Clematis *texensis* and cultivars
Clematis viticella and cultivars (never into wood older than three years)
Hibbertia scandens Snake vine
Hydrangea anomala spp. *petiolaris* Climbing hydrangea
Passiflora spp. Passionfruit vine
Solandra grandiflora Golden chalice vine
Solanum spp. Potato vine
Sollya spp. Bluebell creeper

Table 4.23 Climbers that flower on growth produced the previous season.
Prune after flowering.

Actinidia spp. Kiwifruit, Chinese gooseberry
Aristolochia spp. Aristolochia
Clematis alpina and cultivars
Clematis armandii and cultivars (no pruning except to keep inbounds)
Clematis cirrhosa and cultivars (no pruning except to keep inbounds
Clematis montana and cultivars (no pruning except to keep inbounds)
Gelsemium sempervirens Carolina jasmine
Hardenbergia spp. Native sarsparilla
Jasminum spp. Jasmine
Lonicera spp. Honeysuckle
Mandevilla spp. syn. *Dipladenia* Mandevilla
Tecoma spp. Tecoma
Vitis spp. Grapevine
Wisteria spp. Wisteria

each year to promote fresh growth at the heart of the plant.

Climbers that flower from last year's growth

Grapes, kiwifruit or Chinese gooseberry and wisteria are members of this group but are discussed separately below. These climbers flower from buds formed the previous growing season (see Table 4.23). Those that do not form significant woody structures, such as clematis and jasmine can be pruned to within 30 cm of the ground after flowering once they are established. The growth made after this prune will carry the flowers for the next season.

However, this does not always suit the management of evergreen climbers that have been planted for screening purposes such as jasmine. These vigorous vines can be hedged back after flowering with great success.

Alternatively, remove selected old flowered branches, perhaps one or two a year, to renew the flowering capacity and shorten back any new growth straight after flowering.

The each way bet

Some climbers are as generous as to flower early in the season on last year's growth, as well as blooming late on the current season's growth (see Table 4.24). The dilemma is whether to prune hard before spring and miss the first flowering, or to prune after the first flowering and miss the late summer show. In many ways it is easier not to prune at all. However, this strategy will eventually end in tears; it won't be long before it resembles a large bird nest rather than a purveyor of fine flowers and foliage. Complete decapitation to 30 to 40 cm every few years is one option if the climber will regrow from old wood.

Table 4.24 Climbers that flower on growth produced the previous season as well as current season's growth
Thin older stems in winter and remove spent flowering stems in spring.

Clematis florida
Clematis x jackmanii and cultivars

A more considered approach is to thin older stems in winter and remove spent flowering stems in spring.

Wisteria *Wisteria sinensis* and *W. floribunda*

There can be a certain amount of hysteria when planting wisteria. Perhaps this is due to the truly terrifying (but spectacular) sight a rampantly wild wisteria presents. There is no doubt that an unpruned wisteria is truly beautiful in flower; however, there are few who have the landscape space for such an indulgence. Most situations call for a wall, pergola or out-building to be covered, but not submerged! Prune a wisteria to a framework of branches and some little attention twice a year is all that is needed for a well-behaved vine.

Start by developing a strong trunk from which the major scaffold branches will arise. Select the strongest leader, remove any side growths and shorten the main stem by a third in winter to promote vigorous growth (see Figure 4.89). Continue this regime until the trunk has reached the desired height. Choose some strong stems to form the framework/ scaffold branches and tie them in. Prune these framework branches back to a metre of new growth in winter until they have extended to fill the space available.

In summer there will be streamers of new growth that should be pruned back to two to three buds before it becomes entangled with the rest of the plant. The flowers form at the base of these new shoots.

Once the framework has become established continue to summer prune to reduce vigor

(see page 26, 'When to prune'). Overly vigorous plants result in reluctance to flower or fewer flowers than would be possible. Winter is a good time to deal with tangled congested growth, thin, side growths as well of course as removing dead or diseased wood.

Grapes *Vitis vinifera, V. labrusca* and cultivars

This discussion is not aimed at the commercial grape grower, but the small-scale producer in home or community gardens. As with wisteria, a trunk and framework must be developed before pruning for fruit can begin (see Figure 4.89). There are two general methods of grape pruning depending on the vigor and fruiting habit of the vine. However, fruit is always carried on new seasons growth that has originated from buds formed the previous year (see Figure 4.91).

Spur pruning

Spur-pruned vines have the ability to produce huge harvests. However, the connoisseur may prefer cane pruning that produces a smaller crop with superior quality.

Once the vine's trunk has reached its desired height, cut out the uppermost buds to encourage side growth. Select two subsequent strong growths to form 'arms' or cordons and tie them to the trellis (see Figure 4.89).

If you are wishing to cover a pergola or create a fan select four or five canes depending on the area to be covered. Space these canes evenly and at wide angles to maximise air circulation and reduce fungal problems.

Figure 4.91 Grapes fruit on growth arising from buds that were formed the previous season. Last year's cane has been twisted around the support and the new growth carries the new canes and flower/fruiting buds.

Figure 4.92 Cut back the canes from the recent season to two buds. The growth from these buds will carry next year's crop.

As the season progresses, long canes will grow from the buds on the secured canes. This new growth will carry the fruit. This fruit should not be allowed to mature so that the energy of the plant goes into establishment of the vines' framework, rather than fruit production. Cut off 60% of the flowers and wait two to three years when the vine has established for a reasonable harvest.

In winter, cut out completely any thin, weak or crowded growth, leaving growth that is as least the thickness of a pencil with approximately 15 to 20 cm between them. Prune back the selected canes that were formed that season to two buds. Next season, the growth arising from these buds will carry the year's crop (see Figure 4.92).

In following years repeat the same procedure always pruning to leave a stub of two buds to form next season's growth. See Figure 4.93.

Cane pruning

Some grape varieties do not produce fruiting growth from the basal two buds that are relied on for spur pruning, so cane pruning is the method used. Cultivars such as sultana/Thompson's seedless, sauvignon blanc, chardonnay, cabernet sauvignon and Grenache should be trained in this manner. Cane pruning also yields fewer bunches of higher quality grapes.

This method can also apply to any weak growing vines as it maximises leaf coverage in spring, which in turn strengthens the plant. Much will depend on your climate so seek out some local knowledge.

The same procedure applies for cane-pruned vines for the development of the trunk. When the time comes for developing the horizontally trained 'arms' prune these back

Figure 4.93 Over time, the spurs become gnarled and woody.

to about 25 cm long. In winter select one to three strong canes on each arm, at least pencil thickness and spaced 15 to 20 cm apart. The more canes selected the greater the harvest, but at the cost of quality. Depending on the climate, the more canes selected to produce next year's harvest, the greater the risk of fungal disease. Cultivars such as sauvignon blanc that grow close to the cane may not receive enough light to develop fully when there are many canes wound together. A single cane is best.

Tip prune the cane or canes to eight to 12 buds and wind them around their support. Shorten back a further one to three canes to two buds. These will produce the canes to be selected for fruit bearing growth in two season's time. Prune out the rest of the growth (see Figures 4.94 and 4.95).

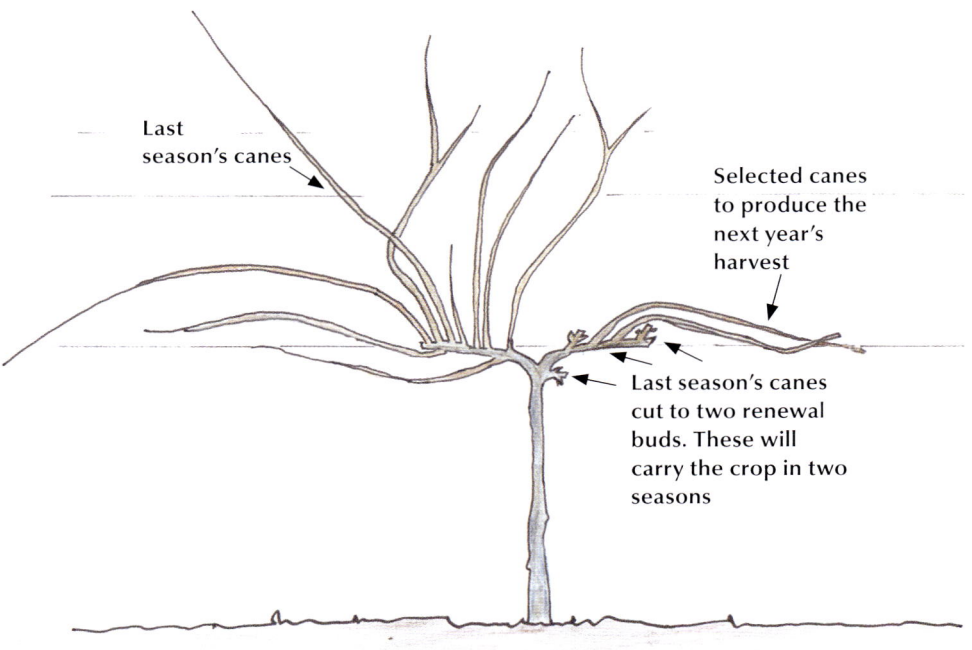

Figure 4.94 Cane pruning.

Figure 4.95 The new fruiting growth arising from last year's cane.

Kiwifruit *Actinidia deliciosa,* syn. *A. chinensis*

Kiwifruit, also known as Chinese gooseberry, is a rampant grower and therefore needs a strong hand to keep it manageable. Like the grape, fruit is borne on new growth arising from a bud formed the previous season. Male and female plants are required for fruit production; the vines vary only in the shape of their flowers. The only time they can be differentiated is when they are flowering, so always prune back the male directly after flowering to avoid confusion. Males flower on the same wood as the female so prune them in the same manner, just earlier. Cut out the flowered wood and shorten back the side growths to produce next season's flowering wood.

Train the vine to a trunk as with wisteria and grapes (see Figure 4.89), selecting two or more stems to act as the vines framework. Always tie kiwifruit stems to their support, never twist them around as they can strangle themselves and sap flow is reduced.

As mentioned before, kiwifruit crop on new growth arising from a bud formed the previous season, so once the framework has been established, select side growths about 30 to 40 cm apart and prune to two buds from the framework. From this point on the pruner must be ruthless. Kiwifruit can be difficult to pollinate so keep the vine open to allow maximum bee access. Plant some bee-attracting plants nearby and even hand pollination will boost fruit set (see Figure 4.96).

Once the trunk and major branches have been established, pruning for fruit can begin.

Summer pruning

The vine will have produced masses of new growth from its main arms. Most of this should be removed, especially the vertical and extremely vigorous. Prune back the fruiting canes (arising from last year's stubs) to two to four buds beyond the set fruit. This will direct the vines energy into fruit development rather than useless vegetative growth. In hot regions leave more buds to shade the fruit; in cooler climes, leave less to allow light in to ripen the fruit. Cut out all other vegetative growth excepting a less vigorous shoot near the fruiting arm. Shoots with horizontal growth and buds packed closely together are the most fruitful. These can be shortened to two to three buds that will provide next year's harvest. Other leafy growth can be removed throughout the summer as it arises.

Winter pruning

Always prune in early to mid winter as the vines tend to 'bleed' their sap later in the

Figure 4.96 Pruning kiwifruit.

dormant season. Prune out fruiting arms that have produced for two or three years; you will be able to see where the fruit has been carried. Select a not too vigorous shoot near the base of the old fruiting wood and prune back to two buds that will produce the next crop. Again, be ruthless with the other growth.

In winter select canes 30 to 40 cm apart and cut them to two buds from the framework – these will carry next year's crop. Prune all the rest out. In summer remove vigorous vertical growth and any cane thinner than a pencil. This should still leave some canes growing from the framework (**1**).

By the next summer the stubs cut in early winter will have flowered and fruited.

Shorten them back to three to four buds from the fruit. Cut out any rampant growth leaving side growths near the base of the fruiting arm (**2**).

Next winter shorten back the growth near the base of the fruiting arm to two buds from the framework and deal with any other over-vigorous crowding growth. Repeat the summer tidy up as shown in step 1 (**3**).

In winter cut out completely the two-year-old fruiting arm and shorten the selected growths off the main framework to two buds as in step 3. If the major fruiting arm has extended to fill the space available, prune it back to just in front of where it started to grow the previous season (**4**).

Note: The male vine, essential for pollination, needs the same regime of pruning. However, the winter pruning should be performed in spring straight after flowering when the male and female can be differentiated.

Clematis

As with any genus that has been highly cultivated, pruning times and frequency vary considerably. Clematis species are listed in the various tables as to the best pruning method and timing; however, there are a few general rules that should be followed.

Pruning at planting

Prune young plants hard to 20 cm to encourage plenty of growth from the base until the plant has a good framework. A 'good' framework is when there are enough established older stems to serve your purpose (see below).

Landscape purpose

When covering a wall or fence, train up to 10 stems and spread them evenly over the surface. The more stems there are, the more flowers. Newer growth can be pruned back to fat buds 15 cm from these stems at the appropriate time (see Figure 4.97).

Clematis that are grown through trees can be trained to carry the vine into the canopy of its host plant. After the initial pruning at planting, select three to four stems and prune them to just below the canopy of the plant that you want it to grow through, up to 2 m from the ground for a tree. This gives the

Figure 4.97 Last season's growth can be pruned back to fat buds 15 cm from the older framework stems at the appropriate time.

climber a 'leg up' so it will flower on the outside of the canopy rather than waste its blooms on the inside of the host plant. When it is time to prune (see Tables 4.22 to 4.24), reach into the host plant and prune hard back to 15 to 20 cm just above the initial pruning point on the slender 'trunks' (see Figure 4.98).

Figure 4.98 Prune clematis hard back to 30 to 20 cm from the ground before the first spring after planting. This will stimulate plenty of new stems.

Most *clematis* are very frost-hardy, however, adjust your pruning time if your climate is extremely cold.

If you are training your clematis through a tree or shrub select some strong stems to grow to the base of the canopy it is to grow through. Prune off new growth to within a few strong buds of these slender trunks at the appropriate time. Secure the main stems/ trunks to their host plant with soft plant ties as clematis stems are brittle and easily damaged. This technique is best attempted with clematis from Tables 4.21 and 4.22.

Clematis are deep-rooted plants so they can be planted reasonably near the trunk of the host plant. The climber's support trunks can be wrapped around the tree trunk, allowing

the growth to be spread through the entire canopy rather than just one side.

Some clematis species are not pruned hard every year as they provide a permanent presence in the landscape and will not rebloom within the same season. *Clematis armandii*, *C. montana* and *C. cirrhosa* are pruned just to keep within their allotted space. The time will come when some renovation will be necessary. There will be plenty of old unproductive wood smothered by newer growth.

To avoid this problem, use a hedge trimmer every two to three years to reduce the bulk of the foliage. If left longer than this there may be no alternative but to replant. However, if they are pruned back to near the base of new wood from an early stage, they will rejuvenate easily without the need for major surgery. Early training to create a well-formed framework does pay off.

> Never prune into wood that is older than three years as it may not resprout.

Flowering time and feeding

Always feed your clematis after it has flowered. The large-flowered clematis such as *C. x jackmanii* and *C. vitcella* cultivars can repeat over the season if they are well fed. Just like modern repeat flowering roses, they will rebloom after a light prune in six to eight weeks.

Pruning weather-damaged plants

The life of a gardener may not be easy, but compared to plants we at least have a choice of location and can avoid bad weather.

Plants vary in their tolerance of wind, sun and frost. Many gardens are made up of plants from diverse climates and there are always some that cannot cope with our increasingly extreme weather conditions. Good plant selection for your area is of course the best place to start; however, there are extreme conditions that test even the hardiest and best-selected plant material.

Wind can be a problem, and pruning to reduce wind loading on a plant is illustrated on page 131, Figure 5.11. The best that can be done is to thin the canopy and ensure that the branches are as stable as possible (see 'Branches', page 19).

Frost and sun damage are heartbreaking. What makes it worse is that it is always best to tolerate the damage until the risk of extreme heat or cold is over before pruning. Enduring the sight of burnt and blasted leaves and stems is unavoidable as pruning will only encourage new growth. Soft young growth is even more susceptible to damage than what was previously killed. Not only that, those crusty wilted leaves and stems are actually protecting the live growth beneath them. They act as an extra layer of insulation between the surviving plant and the extreme weather that may still come. Only when the risks of adverse conditions are over should this depressing mess be pruned back to healthy growth.

Woody plants

When the time is right, after all risk of frost or extreme heat, prune the damaged wood to a healthy bud facing in the direction you want it to grow. Do not prune heavily if it means the removal of live stems. They may look ridiculous, but they could possibly burst into leaf that will enable the plant to recover more quickly (see 'How plants make their food', page 5).

Evergreens that have partial leaf damage (see Figure 4.99) can be left unpruned. The leaves are not attractive, but they are leaves that feed the plant through photosynthesis, thus speeding recovery.

The leaves will grow unevenly through the season, but they will grow, and be overgrown by fresh healthy foliage.

Gardeners in extreme continental climates should also be wary of sun damage on

Figure 4.99 Leave evergreens with partial leaf damage unpruned until after the risk of frost or extreme heat has passed.

previously frost-damaged plants. Stems and trunks that are unprotected by leaves may easily be burnt. In such areas, the use of shade cloth or painting trunks with white water based paint can prevent further damage.

Some plants take the easy road out of trouble by suckering from the base. In the case of grafted plants such as fruit trees (especially citrus), the rootstock will dominate. This will result in the demise of the desired fruiting part of the tree. If all of the wood above the graft and suckers is dead, remove the entire plant and replant with a more tolerant species. If the wood above the suckers/graft is alive, remove the suckers as soon as possible. This will give the desired species grafted to the rootstock the best chance of recovery (see Figures 3.4, 3.5, 3.6, and 3.30, 'Suckers', pages 31 and 46).

Succulents

Succulent plants are rarely troubled by heat but frost can kill the tips of the leaves. Watch these dead parts carefully. If the dead area appears to be increasing it is best to remove them, despite further weather risks. Succulents are very prone to fungal rotting and any necrotic tissue on the plant will invite fungal infestation that may kill the plant completely.

Non-woody or herbaceous plants

Plants without a woody structure, such as many strappy-leaved plants or ornamental plants that are dormant in the winter, will almost always look the most miserable. If they

Figure 4.100 This agapanthus is so ideally suited to the site it is considered an environmental weed. One freak day of 48°C was more than it could cope with. It will recover.

are established plants, however, they have fleshy root systems that act as food storage units that will help them to survive. When the risk of adverse weather conditions are over, they can be pruned back, if necessary to the ground. If strappy leaved plants (monocots) are damaged, the tips of the leaves are most likely to be affected. Remove the dead parts as the leaf will not regrow from the tip (see Figure 4.100).

If herbaceous species are newly planted, they are unlikely to recover. That is just one of the many trials of working with nature. Herbaceous species that are sensitive to frost should be planted in late spring so that they can establish their root system over the growing season. If well cared for, this will make them strong enough to survive the next winter. If not, perhaps you need to rethink your plant selection.

5
FRUIT TREES

Selecting fruit trees

Always buy a grafted fruit tree that is a named cultivar. Seedling trees are an unknown quantity. They will fruit eventually (possibly after many years) but the produce could be not worth waiting for. Seedlings are a bit like children – you never know how they are going to turn out, or know which set of genes they are going to express (see page 21 on seedlings and sexual propagation).

A grafted tree is a clone. The piece of wood (scion) that is grafted onto the rootstock is a piece of the original cultivar. For example, a Beurre Bosc pear tree bought in a nursery today has the aboveground parts identical to the original tree selected in 1807 in Belgium. It is a tree in at least two parts, the scion wood, a bit of Beurre Bosc wood in this case, grafted on to a rootstock (see page 23, 'Asexual propagation', for general principles).

It is best to start with a young tree, a whip or a feathered maiden. These two rather contrasting terms mean an unpruned plant with either just one stem (and pretty whippy they are) or a plant that has developed small side branches (see Figure 5.1).

Rootstocks

Rootstocks provide the root system of your tree, influencing the growth rate and

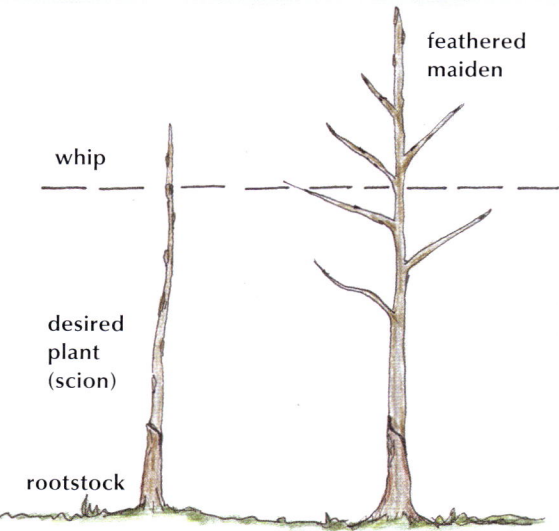

Figure 5.1 When planting your new tree prune to buds/ small branches that are facing in the direction you want growth to occur.

sometimes disease resistance of your fruit tree.

Dwarfing rootstocks reduce the vigour of your tree making them easier to manage by growing more slowly. Generally trees on dwarfing rootstocks fit better into the 21st century lifestyle of less space and less free time. It will not be necessary to chase them with the secateurs, cutting out unwanted vegetative growth. The overall size of the tree is reduced without reducing the size of the fruit. If you chose to espalier your trees you can fit four espaliers in the space taken by one free-standing open vase-trained tree on vigorous rootstock.

If you are attempting to grow a tree outside its preferred conditions (for example, apples and pears on impoverished sand, or citrus on strong alkaline soils), you may be wise to select a tree grown on non-dwarfing stock, as the trees may need as much vigour as possible just to survive successfully. Dwarfing rootstocks by definition have smaller root systems than trees grown on conventional vigorous stocks. The whole point is that they are less vigorous and easier to control; however, the root system is brittle and they will need staking.

There are some dwarf fruit trees that hold Plant Variety Rights (PVR) or Plant Breeding Rights (PBR) which works like a patent on the plant. They have trade names such as 'Ballerina' for some apples, or 'Trixzie' for peaches and nectarines. These are completely different from a non-PBR fruit tree on varying rootstocks. For example, a Granny Smith apple can be grafted onto a M26 rootstock, a semi-dwarfing stock ideal for home gardens making a tree about 2 m in height. It may also be grafted onto a M27 rootstock that is very dwarfing for those who want to train a step-over (see Figure 5.2) to about 1 m to 1.2 m high. A Granny Smith on a M25 rootstock will produce a tree 3 to 4 m high. In each case the actual apple is the same; the height of the tree varies.

Sourcing non-PBR fruit trees on these rootstocks can be problematic. They are available from specialist nurseries and by mail

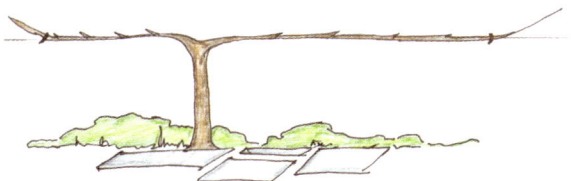

Figure 5.2 A 'step-over' or inclined cordon is usually grafted to a very dwarfing rootstock and should only be attempted with spur-bearing apples and pears, pears being the easier of the two to train.

order. It is always good to order your trees well in advance of planting to ensure that you secure your desired plant. Order in mid summer for winter planting.

Those dwarf fruit trees with PBR, mainly apples and peaches/nectarines, are only available as a patented package. There is a rootstock called 'Trixzie' that produces full-sized nectarines and peaches that are best grown as free-standing trees of 1.5 m high and wide. There are also 'Ballerina' apples that grow as a small column that are ideal for planting at a 45° angle to form a type of Belgian fence (see Figures 5.3 and 5.13).

Free-standing fruit trees

Free-standing fruit trees are for those with a bit of garden space to play with, or perhaps

Figure 5.3 This Belgian fence or *Pallissade Belg* is the most elegant of garden dividers. Plant your trees about 1 m apart. This is best attempted with apples or pears.

Figure 5.4 The only difference between a standard and bush-trained tree is in the length of the trunk.

you just want trees that look 'treeish'. Some fruit trees are best grown as an open vase shape to minimise disease problems, or perhaps you want an orchard of lollipop-shaped trees or an avenue of pyramid-like cones that also bear fruit. Whatever shape you are after, a disciplined construction through judicious pruning and training will make the most of each and every tree. There are three basic shapes: standard/bush, spindle/pyramid and the open vase shape (see Figures 5.4–5.6).

Figure 5.5 The spindle shape is highly productive, but temporary props may be needed to support long branches laden with fruit (right). The pyramid is less formal without the need for training strings with branches that incline more to the vertical (left).

Figure 5.6 The open vase shape suits fruit trees on vigorous rootstocks and for those with more garden space who don't mind using ladders to prune, pick and net. It is ideal for trees prone to fungal attack like apricots.

Remember that the apical bud (see page 3) plays an important role in establishing a framework of scaffold branches, whether it is a delicate espalier or a thumping great open vase tree. The apical bud stands for vegetative growth, wood production, which is exactly what a framework needs! Therefore, allow the apical/terminal buds to extend unmolested until the scaffold/framework branches are as long as needed. Shortening well-spaced side growth only will encourage fruiting wood. Training branches to the horizontal to encourage more fruiting wood will also slow extension growth. When training espaliers allow the ends of the branch to tilt upwards to keep the branch progressing.

Often young trees have branches of varying vigour. One branch is thick and strong, another is rather thin and weedy. Prune back the strongest branches further than the weaker. To encourage the very weakest, leave them with their apical bud intact.

Pruning out too much growth at planting can inhibit establishment. Gauge the vigour of your tree and prune accordingly. Apart from removing very weak wood, competing leaders or crossed branches, the structure can be refined after the tree's first season in the ground. By this time its root system should have established well.

Standard and bush-trained

These shapes are only differentiated by the length of the trunk. A 'standard' tree generally has a much longer trunk before the business of branches and leaves begin (see Figure 5.4). Although undoubtedly elegant, a long trunk topped with a dense canopy carrying fruit should not be attempted on windy or exposed sites and may often need a sturdy stake to secure it in the ground. An unstaked tree may lose its entire canopy if unsupported. A bush tree has a much lower centre of gravity as the canopy is closer to the ground, so complete canopy loss is unlikely. Consider your site and choose accordingly. Depending on the space available vigorous or dwarfing rootstocks may be used; however, for standard trees, dwarfing stock well-staked is recommended.

Start with a feathered maiden or whip (see Figure 5.1), and when established, gradually remove the side branches to the height off the ground desired. This may take a few seasons. Removing too much growth all at once will eliminate the food-producing leaves that speed establishment. Shorten back any side growths that are competing with the central leader. In winter when the uppermost branches have reached the total desired height, cut out the central leader leaving at least two tiers of side branches. This pruning point will be roughly the centre of your ball-shaped canopy (see Figure 5.7).

The side branches should be cut back by approximately a third to encourage branching. These will form the basis of the tree's framework. In spring and summer, any vertical growth should be discouraged and cut to within two buds of the framework branches, and crossing or crowded growth should be removed (see individual entries, pages 143–179, for how to manage specific fruit trees and long-term maintenance pruning).

Spindles and pyramids – a central leader

These two styles are again a variation on a theme (see Figure 5.5). They both have a central leader with side branches being longer near the ground and becoming progressively shorter as they ascend the central trunk. These shapes are best attempted on trees grafted to a dwarfing rootstock. Pyramids are less formal than spindles and require less training.

Spindles are a formal shape akin to espalier, with branches trained to the horizontal to produce the maximum amount of fruit. Strings attached to weights or fastened to a trellis or stake in the ground keep the branches growing horizontally. Soft plant ties attached to baling twine fastened to a lump of wood or a brick is ideal. They can easily be moved out of the way of mowers and then replaced (see Figure 5.5); however, spindle trees that are not trained to a trellis, have long basal branches that can break if they are unsupported. This is especially so when carrying a full crop; temporary props may be needed.

Figure 5.7 Bush or standard-trained.

The pyramid-shaped tree is a less formal affair with branches that incline more to the vertical as they are not restrained by strings.

The branches of any tree trained to a central trunk/leader should never be closer to the trunk than a 45° angle; for spindles an 85 to 90° angle is ideal. Planting distances of 3 m to 4 m is recommended.

Plant your feathered maiden or whip and when established cut away the leader at a metre to 1.2 m high, leaving two sets of side branches to start at between 45 cm to 60 cm from the ground – trim them back by a third of their length. The first summer let your trees just grow, only pruning out very vertical growth growing towards the centre of the tree to two buds/leaves. If you are aiming at a spindle shape use weighted soft strings (that do not cut into the branch) to slowly pull the side branches down to the horizontal (see Figure 5.8). These will become the main framework branches. The next winter cut back the new growth made in summer by about two-thirds to a bud facing downwards, also remove any crossed or overly vertical growth. Tip the central leader by one bud to keep the lower buds alive. Always cut to a bud on the opposite side of the stem from last year's top bud. This will give the appearance of a straight stem in years to come.

Figure 5.9 Open vase training.

Figure 5.8 Spindles and pyramids. After initial pruning, prune new growth in summer to a downward facing bud.

Shorten the side branches by about a third of their length and trim the the central leader to 1 m to 1.2 m. Prune the summer growth by about two-thirds to a downward-facing bud. Pinch out the top of the central leader to a bud facing the opposite way from where it was cut the previous winter. Continue to prune the new growth in summer according to the type of fruit tree you are growing. When the central leader has reached the desired height, cut it back to two buds of the new growth (see individual entries, pages 143–179, for how to manage specific fruit trees and long-term maintenance pruning).

Open vase

The open vase shape is the 'traditional' shape for orchards, producing large trees very open in the centre with side branches arching outwards (see Figure 5.10). As a landscape feature in large grounds, they present the perfect air of tranquility. Be warned, large spaces and muscley friends are needed to lug the ladders required for their maintenance. Trees grafted onto vigorous rootstocks are

ideal for this method, and planting distances of 6 to 8 m are needed.

Plant your feathered maiden or whip and when established prune it back to three healthy side branches. Shorten the branches by about two-thirds always pruning to an outward-facing bud.

FREE-STANDING TREES

- There are three basic shapes for free-standing fruit trees
- Pyramid and spindle trees take up the least space and should be on dwarfing rootstock
- Standard trained trees should not be grown on windy sites
- Open vase training is excellent for trees prone to fungal infections, e.g. apricots
- The more horizontal the branches the more fruitful they are
- Remember that each leaf springs from a growth point/bud/node that is capable of originating a shoot/branch
- Don't cut away too much wood in the early stages of training. The more leaf growth there is the sooner the plant will establish

Figure 5.10 A really fine example of espalier. (Photo courtesy David Glenn, www.lambley.com.au)

As these three branches produce new growth in summer, select the two strongest well-spaced shoots on the original branch to serve as the next limbs. If they are very vigorous, prune them lightly. If you wish them to extend further without branching, leave them alone and cut out any shoots surplus to requirements (see Figure 5.9).

Repeat this process next summer when the year-old shoots will produce another set of shoots. Again select two and prune out the rest. This forms the basic framework for a classic five to six limbed tree. Branch angles from the trunk should be no less than 65°. The branches will eventually take on a weeping shape as the fruit weighs down the new growth (see individual entries, pages 143–179, for how to manage specific fruit trees and long-term maintenance pruning).

Espalier – trees in small spaces

To quote Alan Edmunds, to grow an espalier is:

'To produce a tree that can be trained, pruned, sprayed, psychoanalyzed, exhibited to friends, fed, fondled and filched of its fruit with the least possible expenditure of perspiration and time.'

The art of espalier is possibly the most sophisticated of all pruning and training techniques. It takes a dedication and attention to detail that requires the enthusiast; however, as garden plots are ever decreasing, and the desire to grow our own fruit and have some semblance of self-reliance in food is ever increasing, it is an art worth cultivating.

Espalier is a method of pruning that allows the gardener to produce the maximum amount of fruit in a restricted space (see Figure 5.10). They are easy to pick from, easy to prune and easy to net for birds.

It is the pruning style for gardeners of the 21st century – and it really isn't that difficult.

It has been said that the definition of espalier is as loose as the designs are formal.

They can be called wall trees, espaliers, cordons, palmettes, step-overs and possibly many other unprintable things in the process of training!

An espalier, in the sense that I use the word, means a plant trained to one plane. It is a two-dimensional plant pared back to reveal a skeleton of mother branches well furnished with productive wood. It is a marriage of art and science; ultimate gardening! It allows the maximum amount of air and light to each bud, making the most efficient of fruit trees by eliminating unproductive wood.

Design and building an espalier

Such an elegantly economical tree as an espalier consists of three basic aboveground parts: the rootstock that determines the tree's vigour, the leader, trained to slow its vigour, and mother branches that bear the fruiting branches/spurs (see Figure 5.11).

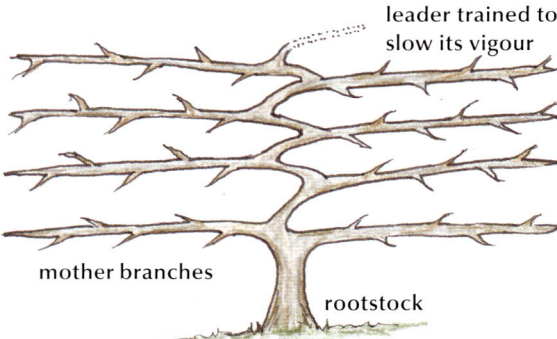

Figure 5.11 This basic horizontal design utilises a bud arising from the mother branch as a leader. The bud/branch is bent to the next training wire to create the next tier of mother branches. This zigzag pattern known as *Palmette Gressant* minimises apical dominance, consequently slowing the sap and reducing the leader's vigour.

Figure 5.12 The fan-trained espalier. Always keep the centre of the fan open and encourage downward growth. The fruiting branches are temporary for peaches, almonds, nectarines, Japanese plums and sour cherries. Sweet cherry and European plum will have more permanent branches as they fruit on long-lived spurs.

The pattern in Figure 5.11 makes it easy to grow a uniform espalier producing fruit from top to bottom. This is best suited to apples, pears and quince. European plums and sweet cherry can also be trained this way but are more suited to fan training (see Figure 5.12).

The 'step-over' or inclined cordon is ideal for lining a path or garden bed, and consists of one long mother branch trained just a metre from the ground. It needs a frame to be trained to as any other espalier; a long piece of old water pipe is a sturdy solution (see Figure 5.2)

Earlier in the book I explained how the cambium (sap, lifeblood) of a tree works, so we know that generally vertical growth produces vigorous vegetative (non-fruiting) wood and horizontal growth produces fruiting wood.

The easiest espalier designs, therefore, focus on growing mother branches as horizontally as possible. It is also true that a central leader (product of an apical bud) will

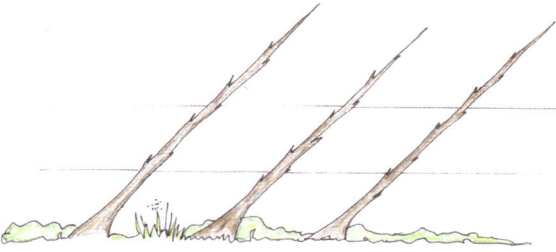

Figure 5.13 'Ballerina' apples make for easily-trained cordons that can be combined to create garden dividers.

continue to favour the uppermost parts of the tree, at the expense of the lower branches. An effective espalier needs to be clothed evenly from top to bottom, so avoiding vertical growth, and a central leader, makes training much easier.

A variation of a central leader, often called a cordon, is successful with fruits that bear on small outgrowths called spurs, such as apples and pears (see Table 5.2). It is always advisable to train these at an angle to minimise apical dominance (see Figure 5.13).

A variant of this method can be combined to create a Belgian fence, perhaps the most elegant of garden dividers (see Figure 5.3). Suit your design to the type of fruit you are growing (see Table 5.1). There is vast scope for creativity when designing your espalier. All you need to do is to keep in mind the natural laws of growth outlined on pages 1–7.

Table 5.1 Fruit trees suitable for espalier training

Horizontal training
Apples, pears, medlars, quince and European plum
Fan training
Japanese plums, sour cherry, almond, nectarine/peach, fig and quince, persimmon

Table 5.2 Apples and pears best grown as free-standing trees
Some apples and pears are not suited to espalier as they fruit on branch tips as well as spurs. If the tips are cut back the tree will have a reduced harvest. The following varieties are best grown as free-standing trees (see pages 126–132).

Apples – Jonathon, Granny Smith, Golden delicious
Pears – Williams (syn. Bartlett)

Support and trellis

Select your tree and ensure you have the cultivar you require, ideally on a dwarfing rootstock (see page 125). If you can choose how to orientate your espalier to the sun, a trellis on open ground running north–south will allow maximum sun and air flow to both sides of the tree. If it is being trained against a wall, choose a north-facing one in the southern hemisphere, or a south-facing wall in the northern hemisphere for maximum fruit.

Once you have decided on your design, erect the trellis or mark out a wall or fence while the passion is upon you.

A free-standing trellis using solid uprights well sunk into the soil, and fencing wire or chain strung at intervals where you want to train your mother branches is the simplest construction. Using lightweight chain instead of wire makes it easier to attach plant ties as they can be anchored on a link, rather than free-wheeling along a wire (see Figure 5.14).

Figure 5.14 Using chain rather than wire makes ties stable.

Reinforcing mesh used in building construction is also effective, if less aesthetic. A bracing crossbeam at the top of the trellis will make the structure sound.

When building a trellis for horizontal designs, place the wires 35 cm to 60 cm apart depending on how closely you want the mother branches to grow. Those gardening in cool areas need distances at the wider end of this spectrum so that the branches can be exposed to the maximum amount of light. Those in warm areas can place the wires more closely.

When training trees as fans, place wires between 30 cm and 40 cm and fasten bamboo rods or wooden dowels onto the wires where you want the branches to be trained (see Figure 5.15). This is ideal for peaches and nectarines (see Table 5.1 for other suitable trees for fan training).

If you want to grow your espalier against a wooden fence, use hooks and wire or attach strong wooden dowels or even metal reinforcing rods instead of wires. A brick wall can have wires strung between masonry plugs at least 10 cm in front of the wall or a heavy-duty trellis can be erected.

Generally, when using dwarfing rootstocks a trellis about 2 m wide and 2 m high is enough to contain one tree. If a more vigorous stock is used a framework of 2.4 m high by 5 m long could be needed. If you want to attempt a Belgian fence, make your trellis as long as required and plant your dwarf trees at 1 m spacings to create the desired effect.

The lowest mother branches can be started as low as 45 cm from the ground or as high as

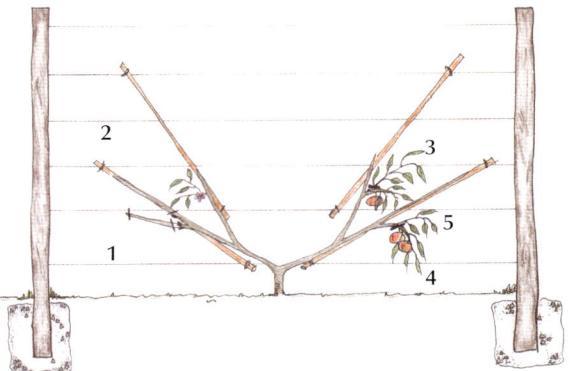

Figure 5.15 All supports need to be solidly built. They need wires strung at 30 cm to 40 cm intervals. For nectarines and peaches pinch out new growth then prune to two buds/leaves in summer (1). First spring growth from the triple bud (2). Next summer, prune out crowded shoots to one bud (3). After harvest prune out the shoot that has fruited leaving a bud (4). The new growth from this bud will grow next spring and carry next season's crop (5).

60 cm. Check your tree and cut it back to buds or branches that are nearly opposite to each other, these will form your first tier of mother branches. The sequence of bud/leaf/branch arrangement is outlined on page 6, so there will be buds on the whip or feathered maiden to suit your purpose within this height range (see Figure 5.1). Again, by observing the bud arrangement, the place to prune in order to create the next tier of mother branches, will be clear. Always cut to a bud facing in the direction you want growth to go.

Managing mother branches or pruning and training for fruit

As your potential mother branches extend, they should be gently trained to the horizontal throughout the summer. Pears appear the most flexible, followed by apples and plums. Peaches and nectarines are more brittle and therefore more suited to fan training.

While the branches are still only partly woody they can be tied into place. The branch will respond almost instantly by trying to grow vertically, uptilting at its ends, the effect of the apical bud (see page 3). Where the branch is (gently) tied there will be a build-up water and nutrients that will feed the buds in that vicinity and encourage fruit buds. Allow the ends of the branches to incline vertically for about 30 cm while you are aiming to get to the desired length of mother branch, and then tie them down again. Growth hardly occurs when a branch is tied to the horizontal so training has to be gradual (see Figure 5.16).

As the mother branches are extending they will start to grow wood shoots (see Figure 5.17). Pinch these wood shoots out at the top when they are about 20 cm to 30 cm long and when they are firmer in two weeks' or a month's time, they can be cut back to two leaves/buds from the mother branch. This will leave a stub of approximately 2 cm to 2.5 cm long.

These stubs will form the basis of future fruiting wood that will no doubt adorn the entire length of branch and make your espalier the envy of the neighbourhood!

Different fruits produce on different wood so the next step depends on what fruit you are trying to grow.

Apples, pears and European plums, sweet cherry

These fruits bear on spurs; knobbly outgrowths (see Figure 5.18).

This talent makes them the easiest of fruit trees to espalier. Having cut your wood shoot

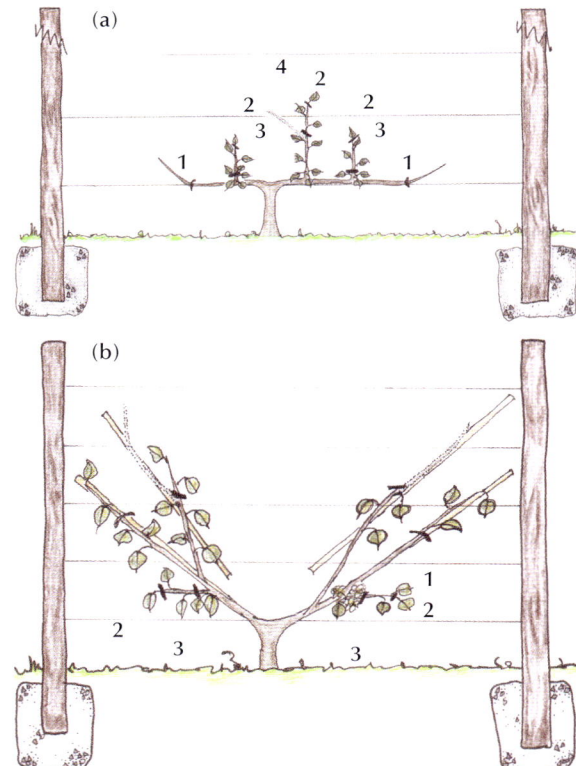

Figure 5.16 Gently tie down the growth that will form the mother branches (1). Pinch out the top of new growth when the wood shoot is 20 cm to 30 cm long (2). When the wood is firmer in two weeks time prune to two leaves/buds from the mother branch (3). Prune to a bud that is growing in the direction required to produce the next tier of mother branches (4).

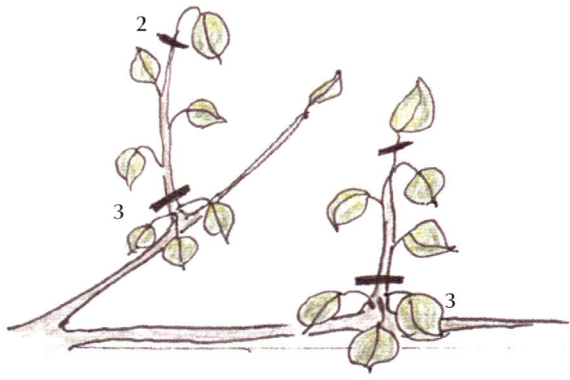

Figure 5.17 Pinch these wood shoots out at the top, then lower in a few weeks time.

Figure 5.18 Left to right: Nashi pear spurs, Macintosh apple spurs, European plum spurs.

to within 2 to 2.5 cm from the branch, the small buds at the base of the stub will grow into fruiting spurs. These may need to be thinned through the years and any wood shoots arising from them should be cut back ruthlessly to within one leaf/bud from the stub. The spurs fruit for many years so it is easy to keep the fruiting growth close to the mother branch.

Aim to have the spurs evenly spaced at 15 to 20 cm apart along the branch – not a cluster together then 60 cm of unproductive wood; a mortal sin in the espalier world!

Some buds along the mother branch may refuse to produce growth. This can be remedied, pinching out the apical buds of branches or by semi-cincturing (see pages 41 and 42).

Peach, nectarine, almond, Japanese plum, sour cherry

These trees fruit on the wood produced the previous spring. They are all strong-growing trees and tend to live fast and die young. Their apical buds race away to the detriment of the axillary buds, so all shoots should be pinched back every 30 cm as a matter of course to redistribute the hormones and cambium flow needed to keep the lower buds alive.

New shoots arising from the branches should be cut back in summer to two buds/leaves after harvest. The buds that develop

ESPALIER FRUIT TREES

- Horizontal growth produces fruit
- Vertical growth produces wood
- Build a secure support/trellis
- Always keep fruiting growth as close as possible to the mother branches
- Prune, pinch out through late spring summer and autumn
- Prune according to the fruiting habit of your tree
- Be aware that plants want to grow vertically not horizontally
- Remove any vertical growth/apical dominance, except when encouraging woody growth for mother branches

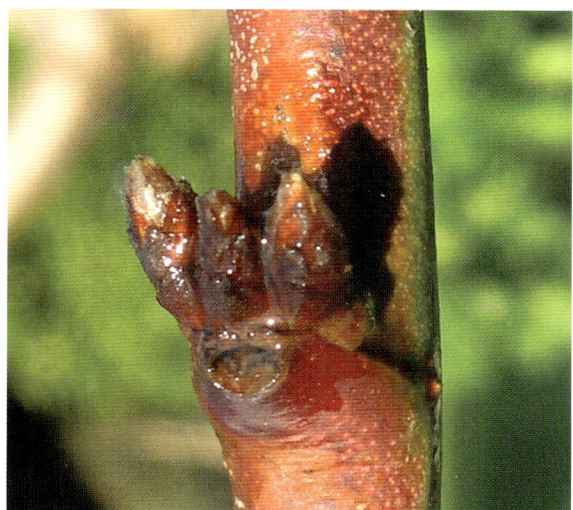

Figure 5.19 The typical triple bud found on peaches and nectarines. The central thin wood bud is flanked by two fat fruit buds.

usually consist of a cluster of two or three buds (see Figure 5.19).

The central bud will produce a wood shoot that will bear next year's crop, while the side buds will produce flowers and hopefully fruit. Shorten the wood shoot to 20 cm in summer, and then after harvest back to its lowest bud. Cut out any extra shoots to prevent overcrowding.

European plum and sweet cherry fruit on spurs like apples and pears, but they respond well to fan training.

Renovating fruit trees

This is a very practical enterprise with the clear objective of fruit production. In the case of a fruit tree, their shape must be open and balanced in order to carry good crops,

and what is more beautiful than a tree full of fruit?

Renovating the scaffold and secondary branches is much the same for all fruit trees; the major difference is in how the fruiting wood is pruned. How you prune the fruiting wood depends on what sort of growth your tree fruits on (see pages 143–179 for the fruit's individual entry).

After removing dead or diseased wood, look for scaffold/structural branches that are the base of the tree's shape. Next, run your eye along the scaffold branches to locate secondary branches and younger stems that are in the right position to form mother branches. Edit these accordingly. Leave the detail of the fruiting wood until last and don't expect large crops after renovation (see Figure 5.20).

This pear tree is very dense with long whippy growth. Long whippy branches will not be strong enough to carry fruit to ripeness; they

Figure 5.20 An overly dense neglected pear tree.

Figure 5.21 The instant removal of dead, broken or diseased wood will get rid of this complicated growth quickly.

Figure 5.22 Clear out the centre of your tree.

will almost certainly collapse and break under the strain. The canopy is so dense that light and air can hardly penetrate, leading to fruit only on the periphery of the canopy and therefore only on about a third of the tree's fruit-carrying capacity. As it is so dense, start by looking inside your tree (see Figure 5.21). Masses of complicated growth can be simplified by the instant removal of dead, broken or diseased wood. This will give you a clearer picture of what there is to prune to create a more open structure of scaffold and secondary branches.

Remove any growth heading towards the centre of you tree (see Figure 5.22).

Crossed branches and vertical growth are the next to go. Prune these off next to their supporting limb (see pages 38–40). The removal of large branches can be

spread over a few seasons to give the plant time to recover. (See Figure 5.23.)

When shortening smaller stems, always prune to an outward- and downward-facing bud or side branch. Try to imagine how that bud or side branch will develop in future years. Will it crowd other branches, or will it form new fruiting growth/limbs with enough space around it for fruit to develop?

Thin the fruiting wood so that all potential buds have space to develop into fruit (see Figure 5.24). Fruiting spurs can become crowded so there is no space for fruit to develop.

At the end of this process there is a clear centre consisting of structural/scaffold and secondary branches supporting the fruiting wood. There are also some branches that could be removed at a later date, possibly in late winter/early spring of the next year when

Figure 5.23 Get rid of crossed branches and vertical growth.

the structure of the tree is easy to see and the plant has recovered from loosing so much photosynthetic material. Bear in mind that pruning in late winter will stimulate vegetative growth (see page 26, 'When to prune'). This is an advantage if you wish to develop new structural or secondary branches. If you wish to restrict the size of your tree, prune again the following summer (see Figure 5.25).

The secondary branches should now be open to the light and air to promote general health and fruiting growth (see Figure 5.26).

The initial renovation has been started and future pruning will further refine its productivity and beauty (see Figure 5.27).

Renovating fruiting wood

When pruning back fruiting wood on nectarines or peaches, cut back the new

Figure 5.24 Thin any fruit-bearing spurs so that fruit has space to develop.

Figure 5.25 A clear centre with a couple of branches that can be removed at a later date.

Figure 5.26 The secondary branches should now be open to promote general health and fruiting growth.

Figure 5.27 The initial renovation has been started and future pruning will further refine its productivity and beauty. The 'sticky' wood that is the remnants of the previous long growth can be cut back to an outward- and downward-facing bud in late winter.

Figure 5.28 Cut nectarines and peaches to a triple bud about 100–150 mm from last year's growth.

season's growth to a triple bud. These triple buds will be made up of both fruiting and vegetative buds. Cut about 100 to 150 mm from last year's growth pruning to an outward- and downward-facing bud (see Figure 5.28).

Figure 5.29 Shorten citrus growth to an outward-facing bud close to the previous seasons fruiting wood.

Figure 5.30 Tidy up the complicated messy old fruiting growth – it will not fruit again, so make room for new growth.

When the major thinning of citrus is done, shorten new citrus growth to an outward-facing bud close to the previous seasons fruiting wood (see Figure 5.29).

Tidy up the complicated messy old fruiting growth – it will not fruit again so make room for new growth (see Figure 5.30).

6

DECIDUOUS FRUIT TREES

Apples *Malus* spp.

Apples are likely to fruit no matter whether you prune them or not. The quality and quantity of fruit is easily improved, however, with a bit of pruning effort. They will also look more elegant. Apples and crabapples have flexible wood, are easy to train and are suited to all fruit tree shapes. All, that is, except the most ambitious of espaliers that are best left to the pears (see 'Espalier', page 131).

Select the shape that fits your landscape purpose and follow the guidelines on how to establish the basic framework of your fruit tree. The rootstock your tree is grafted to will determine the vigor of your tree and govern the success of the style you choose (see 'Rootstocks', page 125).

Codlin moth is a persistent pest in many areas. Unfortunately the use of pheromone traps is only practicable in large orchards. Hanging jars containing a mixture of honey and water from the branches will lure many moths to their death. Three to four jars per tree are effective.

Note: This information is not for commercial orchardists.

Fruiting wood
Spur-fruiting apples

Most apples fruit on spurs. These stout warty growths develop on wood that is two or more years old and can be made up of both flowering buds and wood buds (see Figure 6.1).

Figure 6.1 Most apples flower and fruit from spurs. The fat, furry buds will produce both flowers and leaf growth if the plant is well nourished. Dark pointy buds will grow leaves and new wood only.

Flowering buds are fat and furry while the slender pointy wood buds will produce leaves and if not pruned back, more wood. It is important to remember it is estimated that it takes 40 leaves to nourish one apple! Plenty of leaf ensures a good crop, although some cultivars benefit from some leaf thinning in late summer in order to maximise fruit colour. The wood bud and leaf growth from the fruiting spur invigorates the spur wood leading to further spurs. Shorten back wood growth in summer and it will produce flowering spurs at its base in two year's time (see Figure 6.2).

Over time, the spurs can become overcrowded and will need thinning. Always thin the least fruitful – they may carry only one fat flowering bud and some worn out stubs of spurs. Do not worry too much if the fruitful spurs are further from the scaffold branch, spurs are strong and fruit production is the purpose of the exercise (see Figure 6.26).

Figure 6.2 A side growth with four season's growth. New growth from a wood bud keeps this stem actively growing. Shorten it back in its first summer to develop fruiting buds at its base in two year's time (1). A developing two-year-old bud that will flower/fruit next season. Young flower bearing spurs on three-year-old wood (3). An established spur on old wood (4).

Tip- and spur-bearing apples

Some apples fruit on the tips of short two-year-old side growths. Golden Delicious and Jonathon are two such, with Granny Smith fruiting on both spurs and short side growths. A relaxed attitude to pruning these trees is the best. Leave side growths in place unless they are very long and are disturbing the balance of the tree. In such cases they can be cut back by a quarter of their length.

'They fruit on spurs at my place'

This is the experience of an orchardist in my local area, the Mornington Peninsula in Victoria. In his commercial orchard, trees are managed on a trellis system utilising one central trunk to 3 m high, with up to 22 small branches tied down to the trellis. Using this system, these supposed 'tip-bearing' apples develop spurs. These trees are planted as a whip (or rod) and allowed to branch freely, with all branches tied back to the trellis.

The orchardists work on a ratio of three to one for the trunk caliper compared to the branches; that is, when a branch becomes more than a third of the diameter of the trunk, the branch is removed. One to two branches are removed completely each year in winter and are replaced with selected new growth, again tied down to the trellis.

Maintenance pruning

Once the chosen framework has been achieved, maintenance pruning is all that is needed. Winter pruning will improve vigor and summer pruning will reduce growth (see 'When to prune', page 26). Late winter pruning in frosty areas can delay flowering and therefore minimise frost damage to the blossoms.

Remove strong vertical growth in summer unless it can be trained horizontally to form a replacement branch. Remember that vertical growth produces vegetative growth and horizontal growth produces fruiting wood.

Depending on whether you want to invigorate the tree or dwarf its growth, choose your pruning season and remove any crossed and crowded growth. Always prune to a bud facing in the direction you want it to grow, generally outwards and downward. Dead and diseased wood should be pruned out as soon as you see it.

In free-standing apple trees, removing an old fruiting branch every two to three years will allow you to create a fresh replacement branch. This is best performed in winter. Plenty of new growth will be stimulated. Select one to two strong stems and remove the rest. Weigh or tie this growth down to a horizontal position to encourage fruiting wood (see Figure 6.14).

Fruit thinning

All being well, a snowstorm of blossom (see Figure 6.3) will be followed by fruit set. An apple will 'set' far more fruit than can be brought to harvest so fruit thinning is a must. Not only will fruit just drop off during the season if this job is neglected, the apples that survive the distance will be extremely small.

Orchardists will thin the fruit before it happens; that is, the blossoms will be removed so that the tree doesn't waste its energy on any fruit development at all – only

Figure 6.3 All being well, the pruner will be rewarded with a mass of blossom.

one blossom per cluster will make up the harvest. This is an act of faith that many domestic gardeners are reluctant to make. Know your local climate and supply plenty of bee forage, or even a hive at blossoming to ensure success. If you are growing outside an apple's preferred climate, waiting for the set fruitlets is a safer option.

When you are establishing the framework of a young tree, it is good practice to remove all fruit until the plant is four to five years old. It is better that the tree's energy is channelled into establishment rather than fruit production; a metabolically challenging activity as all pregnant ladies know!

Lack of fruit thinning can also lead to biennial bearing; that is, there are masses of fruit one year and hardly anything the next.

Figure 6.4 For the largest apples, remove all the fruitlets except the central 'king' apple.

Figure 6.5 For medium-sized apples, and to reduce the chance of pests and diseases, remove the central apple and leave two on the extreme sides of the cluster.

Generally this phenomenon is just annoying; however, in some organic systems it is an actual plan. If there is a persistent pest or disease problem, the absence of a yearly crop can reduce the pest/diseases hosted by the fruit. This can make their control much easier.

If the aim is to grow the largest fruit possible remove all the fruitlets except the central or 'king' fruit (see Figure 6.4).

This can feel like the murder of the innocents, but it ensures large fruit and reduces stress on the tree. If a medium-sized apple is desired, allow two fruitlets to remain. The major drawback to this strategy is that as the apples develop, there is little air circulation around them and a site for pests and diseases is created. Removing the central fruitlet and retaining two side fruit can reduce this problem (see Figure 6.5).

If the tree is kept open to the light and is well fed and watered, pests and diseases are kept to a minimum. In some cases only one of the blossoms will be pollinated and the decision is made for you (see Figure 6.6).

A bumper crop should result (see Figure 6.7).

Figure 6.6 Sometimes only one apple in a cluster is pollinated so the choice is made for you.

Figure 6.7 A plentiful harvest. Note that the Granny Smith fruits from both spurs (1) and short side growths (2).

Harvest

When an apple can be removed with a gentle twist they are ready to harvest. Some cultivars ripen over a few weeks such as Gravenstein so picking every few days will yield the best results. Pick them gently so as not to damage the spur wood.

Apricots *Prunus armenica*

Apricots are possibly the most delectable of stone fruits picked straight from the tree. At the time of writing there are no dwarfing rootstocks available so the apricot is a substantial tree, and very beautiful too. They are also prone to a disease called bacterial gummosis, an infection that enters the tree through pruning cuts, so in areas where this is prevalent, great care must be taken (see Figure 6.8).

Figure 6.8 Gummosis is a debilitating disease that enters apricots through pruning cuts. Keep pruning wounds as small as possible and prune in dry weather when the tree is actively growing to avoid infection.

The best form of pruning is light rather than drastic, keeping pruning cuts small rather than hacking off entire branches. Therefore, the best fruit tree shape for apricots is a relaxed open vase system or large pyramid/spindle (see page 126, 'Free standing fruit trees').

In areas where gummosis is not a problem, fan-trained espaliers can be very effective. In such fortunate areas apricots can be pruned hard with impunity as they hold many dormant buds that develop even from very old wood. Their two-dimensional shape makes them easily protected from frost which will devastate the blossom.

Fruiting wood

Apricots fruit on small side growths and spurs developed on one to three-year-old wood (see Figure 6.9).

Figure 6.9 Apricots fruit on small side growths and spurs developed on one- to three-year-old wood.

The spurs are made up of a mass of flower buds, often with a wood bud behind the terminal flower bud. The wood bud/shoot keeps the spur growing strongly, and can be shortened back by a third in mid to late summer so that more spurs develop. Apricots are generally self-pollinating so only one tree is needed for fruit, though some cultivars need cross-pollination. Check with your nursery. Either way another tree in the vicinity can improve fruit set.

Pruning time

Due to the apricot's susceptibility to gummosis and other fungal diseases, pruning time can be critical. Many authorities insist that pruning in mid winter when the populations of fungal spores are low is the best time to avoid infection through pruning; however, it is also damp weather when the tree is dormant and cannot heal itself quickly. Late summer and early autumn pruning, when the tree is actively growing, is preferable. The combination of dry weather and active growth are the best defense against infection, therefore pruning before netting is not only convenient (trees are easier to net), but is the most efficacious. A spray program of Bordeaux mixture at leaf fall and again at bud burst should control fungal infections. *Whenever you prune, always disinfect your secateurs with methylated spirits **at every cut** to prevent spreading disease from one part of the plant to another.*

Maintenance pruning

As usual, all crossed and crowded wood should be removed and over-long side growths shortened. Strong vertical growth usually has the addition of a whorl of sprigs about halfway down the shoot. It is tempting to cut just above these so they develop into fruiting wood. Don't be tempted. Cut below the whorl of growth and you will be rewarded with fruiting side growth much closer to the major branch (see Figure 6.10).

Any diseased wood should be pruned out as soon as you see it, making sure that the secateurs are disinfected before and after.

Aim to thin shoots lightly and remove smaller branches to renew fruiting wood and keep the canopy open to sunlight and air. Cut to a downward- and outward-facing bud to keep the growth low and away from the centre of the tree.

Unpruned trees

Apricots can be successfully left unpruned apart from initial pruning at planting (see pages 129–130). The resulting tree will, of course, be large and may take longer to fruit.

Figure 6.10 Long woody growth often develops small whorls of sprigs. Prune below the sprigs to encourage side growths nearer the parent branch. Pruning before this stem became so long in summer would have prevented this loss of growth.

Left unpruned, the side and vertical growth will eventually become pendulous and fruitful and as the wood is not thinned, growth will be less vigorous. It all depends on what you want your tree to be.

Fruit thinning

To prevent biennial bearing (see 'Apples' page 145), fruit thinning is recommended and it will also ensure larger fruit (see Figure 6.11).

Figure 6.11 Thin fruit to allow enough space for their development – approximately 5 to 10 cm before thinning (a). After thinning (b).

Simply twisting off excess immature fruit leaving a 5 and 10 cm spacing will allow the

Figure 6.12 Apricots should be gently twisted off the tree so fruiting wood and the spurs are not damaged.

fruit room to develop. If this sounds too labour-intensive, whack the branches with a stick to dislodge fruitlets; it also relieves stress in the pruner!

At harvest (see Figure 6.12), twist the fruit gently off the spur or side growth so as not to damage future fruiting wood.

Cherries

Sweet cherries *Prunus avium*

Sweet cherries are generally large trees to 10 m unless they are grafted onto a dwarfing rootstock. Cherries are beloved by birds, so a large tree to be netted requires enormous effort and sometimes grounds for divorce. The solution is, of course, a dwarfing rootstock. The Colt rootstock will produce a cherry from one to two metres high, which is easy to manage in a domestic situation, perfect for fan espaliers or the Spanish bush system (see Figure 6.13). It will also make a neat open vase tree (see page 126, 'Free-standing fruit trees').

The sweet cherry is a naturally upright-shaped tree when young, only spreading when burdened with fruit. If you are planting a sweet cherry on non-dwarfing rootstock remember to allow enough space for it to spread as far as it is tall when mature.

After initial pruning at planting for the chosen shape (see pages 17, 126 and 151), cut back new growth from the selected branches by about three or four buds. This will encourage branching lower in the canopy and prevent the large lengths of bare branch that these trees are prone to (see Figure 6.13).

Figure 6.13 The typical upright growth habit of sweet cherry, here trained as a Spanish bush. It is easy to see where the annual pruning cuts have been made. (Photo courtesy of Bob Magnus, www.woodbridgefruittrees.com.au)

Encourage the branches to spread at a large angle from its trunk/branch by tying down the new shoots with weights or using wooden spreaders (see Figure 6.14).

Weights are easy to manage and easily moved out of the way for the mower. Spreaders are really just a notched stick running from branch to branch to spread the stems. Don't force them too tightly as a stem may split off.

These devices can be used on any tree to spread the branches.

Fruiting wood

Sweet cherries fruit on long-lived spurs that last for many years in much the same way as apples and pears. Most sweet cherries need cross-pollination; however, there are new cultivars that are self-fertile such as Stella, Lapin, Sunburst and Simone.

Figure 6.14 Sweet cherries often have very upright branches. To create more horizontal growth that is more fruitful and a more stable attachment of the branch to the trunk, use spreaders or weights.

Pruning time

Pruning is best performed in summer and autumn to lessen the likelihood of gummosis and silver leaf fungus that can enter through pruning wounds. If you want to promote growth prune in early spring choosing a dry sunny day so wounds will heal quickly. This strategy is not to be advised in very frosty areas as the new growth stimulated by pruning could be burnt by frost.

Maintenance

Once the scaffold branches have been established there is very little pruning to do. Dead, diseased, crowded and crossing growth should be removed together with very vigorous upright growth. Every few years a branch can be pruned out to renew the branch structure.

Cherries do not require fruit thinning, so the major maintenance tasks are netting (or there will be no fruit) and keeping the tree healthy.

Harvest

Sweet cherries are best cut from the tree. This prevents any damage to the long-lived spurs.

Spanish bush training

This method is suitable for both sweet and sour cherries grafted onto the dwarfing Colt rootstock.

This low bush training yields prolific crops by removing the apical dominance of the branches and encouraging fruitful side growths (see page 3, 'Buds – apical and otherwise').

When the tree has reached its optimum height, remove any overly vigorous vertical growth. Each year remove one older branch at its base (see Figure 6.18). Select one or two of the regrowths, and continue to prune it back to 30 cm above the previous cut in the same manner as the initial framework.

By continually cutting out one very vigorous branch each year, the fruiting wood is renewed constantly – and the bush is still only a bit over a metre tall!

Sour cherries *Prunus cerasum*

Sour cherries are a modest size by cherry standards reaching about 4 m with a wide spreading willowy habit. They start to bear

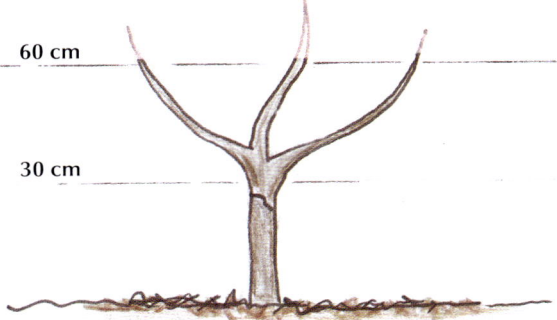

Figure 6.15 Prune back the branches to 30 cm above the graft at planting.

Figure 6.16 At the end of the next growing season it should look something like this.

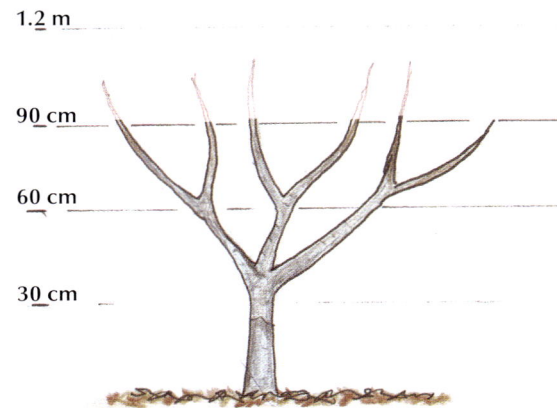

Figure 6.17 In the tree's second winter, prune back the branches 90 cm above the graft; that is, 30 cm above the last cut. Don't even think about what fruiting growth you might be removing. Now is the time to establish a framework, the fruit will be abundant once the foundations for it are laid.

Figure 6.18 Next winter trim back the branches to 30 cm above last years cut. This should make the tree 1.2 m high. By the end of the next growing season the 'bush' should have 15 to 25 branches. Each subsequent year remove one old branch from its base.

early in life and produce the wonderful fruit responsible for Black Forest Cake and cherry vodka. Not as popular with the birds as the sweet cherry; they do, however, need netting. An effective fan espalier, the sour cherry is also suitable for training as an open vase, or a bush/standard or Spanish bush (see page 126, 'Free-standing fruit trees').

Pruning time

Sour cherries get few pruning-induced diseases so pruning time is a matter of choice.

Prune after harvest to thin overcrowded shoots, and in winter to promote growth if desired.

Fruiting wood

Fruit is borne on two-year-old shoots rather than the classic upright spurs of sweet cherries. These slender drooping side shoots remain productive for many years.

Maintenance

Maintenance pruning can be limited to thinning new side growths to about 10 cm apart and removing dead, diseased, crossed or crowded growth. Every few years, a branch can be pruned out to renew the branch structure.

Chestnuts *Castanea sativa*

The majestic chestnut is a magnificent spreading tree best for large gardens and parks. Growing up to 30 m high, it is not a tree to be planted lightly. There are no dwarfing rootstocks for chestnuts, and seed-grown trees are roughly the same height as grafted varieties. Trees are self-pollinating but the presence of another tree will increase yields.

The main reason for pruning is to allow more light into the canopy which will increase nut size. Chestnuts fruit best on one- and two-

Figure 6.19 Chestnuts.

year-old growth that is about the thickness of a pencil, so thinning weak growth will improve yields.

Commercial orchards have occupational health and safety issues to address when they need their crop harvested either by hand or machine. Therefore, they need to keep their trees at about 10 m high. This is usually achieved by removing one or two large branches each year and shortening and thinning side growths in winter.

Whichever approach you take, start by training your chestnut to a single trunk up to 1.5 m high. Select three stems to act as scaffold branches to form a basic open vase shape (see page 130). If you have the luxury of space, practically no pruning is necessary. Prune out any overly vigorous vertical growth that will upset the balance of the tree, and of course remove dead, diseased or crossed growth. Otherwise they can be safely left to themselves.

Chestnuts are also a classic coppicing tree providing high quality timber (see 'Coppicing', page 84).

Figs *Ficus carica*

Figs are one of the most luxurious of fruit and yet are easy to grow in any Mediterranean climate. They are very pliable, and can be trained as espaliers, open vase and bush shapes. They can also be pruned as a low bush to fit into a large pot for frosty areas as they are easily covered or moved to prevent damage (see pages 128 and 129).

Figs are grown from cuttings, and as there are no dwarfing rootstocks available for them,

Figure 6.20 A fig's main crop is always more prolific and of better quality than the breba crop.

they can grow into large trees. *Never root-prune a fig in the ground to restrict its size; it will produce a forest of suckers!* When planting your tree, line the edges of the generous planting hole with concrete slabs or bricks. This will restrict the root system and therefore the size of the tree. It will also make the tree more fruitful.

The milky sap from a figs stems or fruit can cause severe skin irritation. Long sleeves and gloves are the best solution.

Fruiting wood

Some figs produce two crops a year in temperate and sub-tropical climates, while others produce only one (see Table 6.1). The breba or first crop is produced in summer, and is carried on the tips of the previous season's growth (see Figure 6.20). The second or main crop is borne at the base of the current season's growth. This will of course influence how you prune your tree.

The breba crop is often sparse, but the joy of having figs around mid summer can make it worthwhile. Any pruning into the last year's

wood in winter will of course remove any potential first crop. A compromise can be reached where some of the previous season's growth is removed to promote the new growth for the main crop, and some left alone to carry the breba crop. If you have a variety that has no early crop, or you don't want to be bothered with it, winter pruning will produce more new growth to carry the main crop.

Pruning time

Winter is the best time to prune so that new growth is stimulated to carry the main crop; however, leave some of the previous season's growth if you are planning on a breba crop (see above). *When pruning figs always shorten stems back to a bud/node – do not leave a stub that can rot the whole stem* (see Figure 3.20, page 38). In frost-prone areas, summer pruning will allow the subsequent new growth time to harden off before winter. In such areas don't prune in winter unless the tree can be protected – the new spring growth will be burnt off.

In order to make the most of the early fruit or breba crop, pinch back the new season's growth in spring to help it develop. In summer shorten back the new shoots as this will speed the maturation of the main crop in areas with shorter summers. Never remove too much leaf as the bark is prone to sunburn. Some white acrylic paint on the major branches will serve as effective sunscreen. In frosty climates any major pruning is best done in summer.

Maintenance

Established open vase trees may only need pruning once every three years or so. Always remove dead, diseased (rare in figs) or very crowded and crossed growth. Modify your pruning depending on whether you are aiming for two crops a year or only one (see above). Pruning in summer will help to restrict growth, while winter pruning will rejuvenate fruiting wood.

Harvest

Harvest your figs when the neck of the fruit starts to wilt and the fruit droops from the tree. A ripe fig should not exude any milky sap.

Figs in pots, and keeping them small

Figs sucker freely from their roots and to maintain the aesthetic shape of a tree they should be removed. This habit, however, can be utilised for growing figs as a coppice-style plant that can be easily covered for frost protection, or just fitting them into a smaller space.

The fig will not have a formal trunk but will be encouraged to sucker. When the plant has established after three years at least, any wood that is two years old or older can be cut down to the ground or to its base on a short trunk. There should be one-year-old stems remaining that can be thinned if necessary. These will carry any breba crop and the later main crop on their new spring growth. Maintain this system by consistently removing wood that is two years old and thinning the one-year-old wood.

Figs on trellis

Figs are generally trained as a fan espalier (see Figure 5.16b, 'Espalier', page 135). Growing them flat against a wall can make fruiting more reliable in cooler climates, due

Table 6.1 Fig cultivars that bear no or a very light breba crop.

In very cold regions the breba or first crop of any fig will not eventuate due to frost damage of the previous season's growth.

Note: A single cultivar can be known under varying names

Adriatic, syn. White Adriatic, Strawberry fig, Verdone
Brown Turkey, syn. Aubique Noir, Negro Largo, San Piero
Kadota, syn. Dottatto, Florentine, Whote Kadota
Panachee, syn. Striped Tiger, Tiger

to the radiated heat from the wall. It also makes them easier to protect from frost.

The Japanese treat them more as a grapevine with horizontal fruiting arms. Four main fruiting arms are established from a trunk. The new growth in spring grows up and bears a main crop. At the end of the season after harvest, these stems are cut back to two buds which will produce next year's main crop. This is exactly like spur-pruned grapevines (see page 114). These spurs may need thinning, as figs are blessed with many dormant buds that will spring into being with this heavy pruning technique.

Hazelnuts *Corylus avellana*

Hazelnuts, filberts or cobs are shrubby trees that love to sucker. In fact, they are the archetypal coppicing plant providing excellent wood for all manner of woodworking or craft activities (see page 84, 'Coppicing'). Unfortunately hazelnut production and coppicing are mutually exclusive. Good harvests are only possible when the female flowers receive sufficient light, a commodity conspicuous by its absence in close-growing hazel wands.

At the time of writing, the Turkish hazel *Corylus colurna* – a non-suckering hazel – is being trialled, but with limited success. So hazels are grown on their own roots and continue to sucker freely – certainly an opportunity to increase yours or others stock of hazelnut trees, and a constant source of maintenance (see Figure 3.30, page 46). In order to get maximum light into the canopy the open vase tree shape is recommended (see page 130, 'Free-standing fruit trees').

At planting, some sources recommend removing dormant buds. Disbudding the upper root system and on the proposed trunk can prevent suckering. Use a sharp knife to cut/scoop out the bud completely (see 'Disbudding', page 43).

Fruiting wood

Hazels are monoecious, meaning that they carry separate male and female flowers on the same plant (see Figure 6.21). The quietly elegant male catkins appear first followed by

Figure 6.21 Hazelnut's male flowers (catkins, pictured) are carried on the ends of branches. The insignificant female flowers are borne at the base of one-year-old growth.

the tiny female flowers. Hazels are self-sterile so they need a compatible companion tree to fruit. If you are planning a large plantation allow one pollinator to four or five good cropping cultivars. A little research will prevent disappointment.

The female flowers are borne on strong shoots about 20 to 30 cm long formed the previous year. There are usually plenty of males about but female flowers can be elusive, so maximise the harvest by encouraging this wood.

Male flower production can be adversely affected by heat. The dormant buds can be killed by temperatures over 21°C. Remember that the temperature under a hazel leaf will be lower than the general air temperature, but it will be sound practice to plant your hazels away from the hot western sun in warm areas.

Pruning time

Hazels can be pruned in autumn after harvest, removing stems that have already borne nuts, but major pruning should be carried out in late winter to spring after the catkins have fallen.

It is an English habit to 'brut' their hazels in summer. This involves breaking by hand, but not severing, the vigorous new growth to promote the production of female flowers the following year. The semi-broken stems are not removed until winter. I have no knowledge of its efficacy, but imagine that the interruption of the flow of the cambium (see page 1) would slow its flow and promote fruiting wood, much like the practice of espalier.

Maintenance

Once a short trunk and scaffold branches have been established, hazels can be left to their own devices for a few years. The aim is to supply enough one-year-old wood to crop in the coming season, remove any tangled or crowded growth and to maintain plenty of light penetration through the canopy. Suckers should be removed as soon as they are noticed; they will block light and crowd the tree.

Medlar *Mespilus germanica*

Medlars are one of the most beautiful of fruit trees with the additional attraction of yielding fruit in late autumn into winter (Figure 6.22). Establishing a good framework of scaffold branches is all that is necessary for a successful tree (see Figure 4.3 for corrective or guiding pruning at planting).

Medlars fruit on the tips of one-year-old wood as well as spurs that can be happily left to themselves. As a fruit they are not yet fashionable, needing to be bletted, or let to go very soft, some may say rotten. Think of the flesh as looking like caramel pudding.

Figure 6.22 Medlars. A not-yet fashionable fruit on a graceful tree. (Photo Norwood Industries Pty Ltd)

Harvest your medlars after the first frost if you have them, or when they can be removed by a gentle twist. Lay them crown-side down without touching each other and await developments.

Mulberry *Morus nigra, M. rubra, M. alba, M. macroura*

Whether your mulberry is black, red, white or the elongated Shartoot, they are undeniably graceful shade trees with a naturally twisted and elegant form (see Figure 6.23). Generally they grow into large trees but there are dwarf cultivars available – as yet not widely.

Most mulberries are large, reaching up to 10 m in time and apart from pruning at

Figure 6.23 The black mulberry, *Morus nigra*, grows into a beautiful large gnarled tree. There are also dwarf cultivars that are now availab e.

planting (see pages 17 and 126). There is little else to be done other than the removal of dead, diseased or crowded wood. Netting them when in fruit is usually not viable given their size but fortunately they are so prolific that there is more than enough for both people and birds.

Fruiting wood

Mulberries fruit in the leaf axils of actively growing shoots and also on spurs from older wood. As long as the tree is healthy it should fruit with or without pruning. Spur development can be encouraged by shortening back side growths after fruiting.

Pruning time

Prune directly after harvest if you need to prune at all. Some cultivars with *M. rubra* in their breeding will develop the strange habit of trying to produce fruit in mid-winter if they are pruned at any other time.

As for all other mulberries excepting the dwarf, prune in winter for growth and summer to restrict growth.

Dwarf mulberries

The dwarf cultivar of *Morus nigra* does need regular pruning after it has fruited. It is grown as a shrub with foliage to the ground. After harvest, cut back the plant by one-third clearing overcrowded wood as you go. As a shrub it will be multi-stemmed; cut out one major branch every few years to renew the fruiting wood.

Figure 6.24 The delectable nectarine.

Nectarines, peaches, peacharines and almonds *Prunus persica* var. *nectarine*, *P. persica*, *P. dulcis*

Peaches, nectarines and almonds are all pruned in exactly the same manner. Nectarines are a variety of peach, the major difference being the lack of fuzz. Almonds, although a different species, fruits on the same wood and benefits from the same treatment. Such delectable fruit needs plenty of air and sunshine to develop so they are best trained to an open vase or a fan espalier (see pages 132–134). In commercial orchards, they can be trained to a central trunk planted at an oblique angle with their side branches trained at a 45° angle from the trunk, much like a Belgian fence without the initial trunk (see pages 126 and 133). A bush shape is not as effective as having a naturally spreading habit, they tend to turn into an umbrella shape with the best fruit produced at the top of the tree rather than within reach of the picker. Light and sunshine on the fruit/nuts is also less likely.

When establishing the framework of the tree, allow the apical or terminal buds of the selected side branches free rein to develop. Select and prune for the scaffold branches in winter, pruning out really weak stems; they are unlikely to succeed even when attempting to equalise their vigour (see page 127).

Fruiting wood

The best flowers and fruit are produced on one-year-old wood, that is, growth that was made the previous growing season (see Figure 5.28, page 140). For this reason, these trees need quite heavy pruning once they have established in order to renew the fruiting wood. Peaches and nectarines have the ability to produce new growth from dormant buds so new growth is easily stimulated by pruning.

Pruning time

After the initial prune, thin out the new growth in summer. In autumn shorten two-year-old wood, which has already fruited, down to two buds or some strong side growths produced that season.

In very dry areas where no summer watering is available, the late spring early summer growth can be limited. Shorten back side stems at flowering to stimulate growth while water is available. Some of the harvest will be compromised, but this method does ensure the production of future fruiting wood.

Nectarines and peaches need between 200 and 1200 hours of chilling to produce a crop.

In very warm areas where only the low chill cultivars are successful, prune the trees directly after harvest.

Maintenance

Thinning in summer is a task that should not be omitted as light and air need to penetrate the canopy freely so remove strong vertical growths from the centre of the tree. The new long feathery stems need to be shortened back to strong side growths and weak new growth can be removed altogether. Allow a handspan between these new side-shoots, selecting those that are the thickness of a pencil where possible. They will bear next year's crop. Drooping and very short growth can be removed as they rarely produce enough leaves to feed their fruit. Be most severe on the new growth at the top of the tree so that it does not shade the lower.

In autumn cut back the stems that have produced fruit to two buds from the major branch. This will produce new growth in spring that will fruit in two season's time. Once the tree is established remove at least one major branch annually in winter to renew fruiting wood.

Deal with dead or diseased wood as soon as you see it.

Fruit thinning

Peaches and nectarines produce large heavy fruit and so fruit must be thinned so as not to over burden the branches (see Figure 6.25).

Thin the fruit so that they are about 10 cm apart. Almonds need no thinning at all.

Figure 6.25 Lack of fruit thinning and an unsupported branch led to this catastrophe.

Pears *Pyrus* spp.

Pears, be they European or Asian, are vigorous and easy to train. The nashi or Asian pear differs only in its harvest techniques, the word 'nashi' meaning 'pear' in Japanese. So whether they bear buttery and aromatic or firm and crunchy fruit, pears are a beautiful addition to any landscape.

Feel free to choose any tree shape that you fancy – pears are very accommodating. They

Figure 6.26 The easy-going pear.

Figure 6.27 Current season's growth cut back to two buds in summer (1). One-year-old wood forming spurs to flower and fruit next season (2). Three-year-old wood with new one-year-old spurs (3). Old growth with well-developed spurs (4).

make perfect espaliers and can be grown as pyramids, spindles, bush or open vase systems (see page 126, 'Free-standing fruit trees', and page 131, 'Espalier'). Whatever style you choose, aim for strong wide branch angles when selecting which side growths to use for scaffold branches. Pears are prone to vigorous vertical growth that is shy to fruit, so use spreader bars or weights to encourage them to grow more horizontally (see Figure 6.14).

Pears need another pear to cross-pollinate with, so ensure that you have compatible cultivars. The 20th century nashi pear can be self-pollinating but bears better crops in the presence of another.

Buying your pear on a dwarfing quince rootstock will limit its height to 2–3 m, whereas those grafted to pear rootstocks make magnificent trees to 6 m high; a challenge to pick. If you are training a standard or bush pear, cut out the central trunk or leader in summer rather than winter to slow its growth, to minimise the apical dominance, and reduce regrowth (see page 3, 'Apical buds').

Fruiting wood

Just like apples, pears fruit on long-lived spurs produced on two-year-old wood, and their growth is managed in much the same way (see Figure 6.27).

As mentioned earlier, pears are prone to producing vigorous vertical growth that should be weighed or tied down to slow the flow of the cambium and encourage spur development.

Maintenance

Once the framework has been established, maintenance pruning consists of tying down new growth, thinning side growths to a handspan apart, and removing weak or spindly wood. Young trees persist in their production of vigorous vertical growth that, if

not useful for framework building, should be pulled off so as not to unbalance the desired structure of the tree (see page 46, 'Suckers'). As the tree ages, spurs will need thinning to allow space for good-sized fruit to develop (see Figure 6.28).

The spur on the left of Figure 6.27 has one spent spur and is crowding the one above it. The right hand spur is overly long with just one bud. The top spur has been shortened to a bud facing away from the main stem.

Keep the branch invigorated by maintaining, but continually shortening, a wood bud at the extremity of the branch. When branches become too long, cut them back to a strong side shoot or remove them altogether. In free-standing fruit trees, remove an older branch every few years to renew the fruiting wood. Espaliers need more specialised attention as they have permanent mother branches (see page 131, 'Espalier'). As with all trees, remove dead, diseased, crossing or crowded growth.

Fireblight

Thanks to the Australian Quarantine Service, we are mercifully free from this dreadful disease. However, there are many countries that are plagued by it. In such areas, avoid heavy pruning that will stimulate vigorous new growth – its favorite victim. Remove affected shoots to 15 cm beyond the blight as soon as possible and remember to sterilise all tools thoroughly between cuts.

Harvest

European pears should always be picked before they are fully ripe. As soon as they

Figure 6.28 Older pear and apple trees will need their spurs thinned.

turn light green, pick one and if the seeds are dark they are ready to harvest. They store well in cool temperatures and improve in flavour. Nashi pears should be allowed to ripen on the tree.

Persimmon *Diospyros kaki*

Some may be reserved about the fruit but no-one would dispute the beauty of the tree. The rich autumn foliage and deep orange fruits that dangle from bare branches make this the most ornamental of fruit trees. There are both astringent and non-astringent cultivars, but their fruiting habits and pruning requirements are the same.

Persimmons have very fragile wood that is vulnerable to wind and heavy fruit loads, so the establishment of a strong framework of branches at an early stage is essential.

They are well suited to the pyramid or spindle fruit tree shapes with wide branch angles to support the brittle branches. The open vase method should be reserved for cooler areas; persimmons are prone to sunburn. In commercial orchards they are grown on a trellis.

Persimmons can be hard to establish so minimise root disturbance at planting. Feed the tree well in spring as this is often the only season of growth for the persimmon. Strong healthy growth from the beginning will establish the tree quickly.

Fruiting wood

Persimmons fruit from new growth that arises from the last few buds of the previous season's growth (see Figure 6.29).

Figure 6.29 Persimmons produce fruit from the current season's growth (1) arising from the previous season's growth (2).

This means that if you shorten the new growths produced last season, you will be removing all the fruiting wood. The solution is to leave some new growth unpruned and prune others back to within a few buds of the scaffold branch to produce fruiting wood in two season's time. This will not only moderate the crop but will prevent the persimmon's tendency to biennial bearing. There is the option of not pruning at all; however, the resulting long willowy growth will be easily snapped off by the wind, and the fruit will get smaller and further away from the scaffold branches.

Pruning time

Pruning for a stable framework can be done in winter to take advantage of the major spring flush of growth. Pruning out growth that has already fruited can be done at harvest when the fruit is picked.

Maintenance

In winter, prune back to a few buds some of the newest growth and leave the rest. Any wood that has already fruited should also be dealt with if it was not removed at harvest. Get rid of dead, diseased, crowded or crossing growth and persist in encouraging strong wide branch angles in the main framework. The use of weights and spreaders can help train growth more horizontally (see Figure 6.14).

Harvest

Fruit should be harvested when it is still firm, whether you have an astringent or non-astringent persimmon. When it appears that the fruit's skin has reached its deepest colour (experience is the only guide) pick your fruit. The astringent cultivars will need to be bletted. That is, they will be so soft as to be squishy. Lay the fruit out, green calyx down, and wait for the super-soft texture to arrive in a few days; a ripe apple next to them in a paper bag will hasten the proccess. Make sure they are really soft before eating as an astringent persimmon that is not completely bletted is a ghastly experience, though perhaps amusing for others to watch!

Non-astringent cultivars can be eaten straight off the tree, but their flavour improves with bletting.

Pistachio *Pistacia vera*

Pistachios are shrubby trees that need the planting of two trees to produce nuts. They are dioecious which means male and female flowers are produced on different plants. They also benefit from good training when young, the open vase system being the most effective as it minimises shading (see page

Figure 6.30 Pistachios are shrubby trees that need the planting of two trees, a male and a female, to produce nuts.

126, 'Free-standing fruit trees'). In winter, shorten vigorous shoots to produce side growths about every 30 cm, these side growths will become the scaffold branches.

Once the framework is in place, prune in winter to produce vegetative growth. This growth when shortened the next year will produce the flower bearing side growths. Pistachios fruit on these new side growths that arise from wood produced the previous season. They are very prone to biennial bearing so balance the vegetative and flowering stages on your tree to create regular harvests. Encourage some strong vegetative growth and shorten back last year's growth for flower-bearing wood each year. Removing a large branch every few years should renew the wood and keep the tree more compact.

Plums *Prunus domestica, P. salicina*

There are two types of popular plum. The European plums have yellow flesh and

Figure 6.31 European plums can develop elegant weeping habit, but prune to an outward- and upward-facing bud to keep the branches clear of the ground.

include greengages, prunes and general desert plums (see Figure 6.31). The Japanese plums are also yellow-fleshed but have a group called blood plums with deep red flesh. The Japanese plums are a larger tree, flower earlier and have different pruning requirements to the European. The Europeans are easier to fan espalier, while both do well trained to pyramid, spindle and open vase forms (see page 126, 'Free-standing fruit trees'). There are a few self-fertile plums but most need a cross-pollinator. Check that you have compatible cultivars.

Fruiting wood

European plums *P. domestica*

These plums fruit on relatively long-lived spurs that develop on two-year-old wood much like apples and pears (see Figure 5.28,

page 140). Shortening back side growths can encourage their development.

Japanese plums *P. salicina*

Japanese plums fruit best from one-year-old wood as well as producing short-lived spurs. Fruiting buds arise from the previous season's wood so fresh growth each year should be encouraged to carry the next year's crop. Shorten wood that has fruited after harvest when it can be cut back to a strong side shoot of new growth. These growths should be thinned to prevent overcrowding. Prune back long willowy stems by about a third otherwise they will be prone to weather damage.

Pruning time

All plums are best pruned in summer to minimise the incidence of disease. After harvest is the obvious time, but vigorous unwanted growth can be tamed or removed before netting. Japanese plums can have their two-year-old wood shortened by at least a third just after harvest producing new growth that will fruit the next year.

Maintenance

Once established, both these plums need minimal pruning. Prune to correct their shape, remove dead, diseased crossed or crowded growth and to renew fruiting wood.

Always keep them open to the light and air by removing crowded growth. Often the European plum develops a weeping habit so that when shortening new growth or larger branches, cut to an outward- and upward-facing bud or side growth. This will prevent the branches dragging on the ground. Wood buds are generated mainly on one-year-old wood that can be shortened to encourage spur growth after harvest or left to develop a replacement branch. In espaliered trees shorten side growth to six leaves in spring and back to three after harvest.

Fruit thinning

Almost all plums need fruit thinning to reduce the burden of the crop. The European plum is slightly smaller than the Japanese so thin to 5 to 7 cm spacings. Japanese plums are best at 7 to 10 cm spacings. Thinning by hand can be tedious; hitting the branches with a rubber hose-tipped stick is much more satisfying, if less accurate.

Harvest

Plums are best straight from the tree. When they are fully coloured for their cultivar and showing just a touch of softness, gently twist them off.

Pomegranate *Punica granatum*

Pomegranates may not be an essential food but their gnarled beauty and fantastic fruits make them a treat for the landscape as well as the table. This shrubby tree, deciduous in cool climates but evergreen in warmer parts, adapts willingly to many soil types. They will even survive very frosty areas. In regions where the temperature does not go below −5 to 7°C, it is best trained to a short trunk with four or five main branches. A multi-stemmed tree is the best approach in really cold areas because if one trunk is killed by frost, there are others left to continue.

Figure 6.32 Pomegranates flower and fruit on short shoots or spurs near the ends of branches.

Pomegranates sucker freely which can be an advantage in very cold climates if the shrub is grown on its own roots, but generally it is just a nuisance, robbing the main plant of vigour. Remove all suckers as soon as you see them (see page 46, 'Suckers'). Always purchase a named variety as they vary widely in quality and do not come true from seed. The dwarf pomegranate *P. granatum nana* is truly revolting to eat, no matter how ornamental.

Fruiting wood

Pomegranates flower and fruit on short shoots or spurs near the ends of branches. These remain productive for three to four years, after which time they should be cut to a younger side branch to renew the fruiting wood.

Pruning time

Winter is a good time to assess the shape of your plant and perhaps remove a very old branch that is no longer productive. This will stimulate new growth. Select the most robust as a replacement branch. Pruning the four- to five-year-old wood back to a side shoot can be done at harvest.

Maintenance

Pomegranates need very little attention apart from clearing tangled wood, a bit of branch renewal encouraging new growth, and by removing old fruiting wood. The major task is to control sucker growth as described on page 46.

Harvest

Always harvest the fruit before it splits, as the fruit will rot rapidly. Pick when they start to redden in late autumn and before any heavy rain as this promotes splitting. Handle pomegranates gently as they bruise easily despite their firm skin. The unbroken and unbruised fruit will store for a month or more – some say the flavour even improves.

Quince *Cydonia oblonga*

There is nothing quite as opulent as a quince in flower or fruit. Unlike many fruit trees their blossoms appear singly so they are of generous proportions (see Figure 6.33). Quinces are self-fertile and are one of the easiest and most low-maintenance trees to grow. They can be trained as a fan espalier, low-branching open vase or a pyramid or spindle (see page 126). They are charming as an informal tree, however, with five to six main branches originating about 20 to 30 cm from the ground. Shortening back the main stems by about two-thirds for the first two years of growth and keeping the centre clear will provide a stable framework from which to fruit.

Figure 6.33 Quince blossoms are bourne singly making them both large and graceful.

Fruiting wood

Quinces fruit on the current season's growth from a fat elongated bourse functioning a little like a spur that carries both fruit and wood buds (see Figure 6.34).

The new growth from this structure carries the blossom (see Figure 6.35).

Pruning time and maintenance

Winter is a good time to see any crowded or crossing growth and pruning at this time will

Figure 6.34 The growth that produced the just harvested fruit and that will produce next season's bourse (1). The bourse from which the growth was developed (2).

Figure 6.35 The new growth from this structure (bourse) carries the blossom.

stimulate vegetative growth if that is what is desired. In summer, shorten back any strong vertical growth and eliminate spindly unfruitful branchlets.

Often an established quince will throw up a long whippy growth. It will ultimately fruit, but its distance from the main branches will mean that the fruit will be whipped about in the wind damaging both fruit and other branches. Summer is the perfect time to

Figure 6.36 Ripe quinces have an unfortgettable fragrance.

shorten these growths back to three or four buds or remove them altogether if they are crowding the canopy. Light and air through the canopy are essential.

Suckers can be a problem so watch out and deal with them sooner than later (see page 46).

Harvest

When the great globular fruits start to turn yellow and have a discernable fragrance, the moment has come to harvest (see Figure 6.36).

Even though quinces appear to be rock hard, they do bruise surprisingly easily which affects their storage capacity. Be gentle.

Walnuts *Juglans regia*

The English or Persian walnut is an enormous tree that grows to a neat 10 × 10 m tree, so probably not for the suburban garden. The space it takes up is considerable helped along by a secretion from its roots called juglone. This substance effectively retards the growth of other plants so that the walnuts take up the aboveground space, as well as colonising below ground to deter root competition. Walnuts are ruthless.

This delicious nut can be self-fertile but the best crops result from cross-pollination from another cultivar. Not just any other walnut will do, so check that you plant compatible cultivars.

Walnuts have a deep taproot, and as they are often sold bare-rooted, this is often been damaged when lifted from the nursery field.

Figure 6.37 This delicious nut can be self-fertile but the best crops result from cross-pollination from another cultivar.

Fruiting wood

Old cultivars of walnut flower and produce nuts at the tips of their branches. The male catkins are borne on last season's wood, while the females arise from the current season's growth. In such trees no pruning is necessary to promote this wood so keeping the tree open to light and air will allow each nut to have its place in the sun to develop fully (see 'Maintenance', below).

New cultivars that are the mainstay of the Californian industry, bear nuts on the tips of branches as well as on one-year-old side growths. These will need thinning regularly so that the canopy does not become crowded.

Maintenance

Walnuts are prone to fungal diseases and it is important to keep the crown of the tree open to air and light for disease prevention as well as optimum crops. Again, depending on your climate, reducing the canopy can lead to sunburn of both the bark and the nuts. Know your climate and prune accordingly.

Modern cultivars will need their laterals thinned regularly to promote more side growths and also to open the tree to light. Such trees should also have at least one limb removed each year once established.

Harvesting

The outer shell of the nuts will split and the nut drop to the ground when harvest is imminent. Knock a few more off the branches and taste to see if they are ready. If so, harvest them before the wildlife finds them.

The root system on such trees is tiny compared to the aboveground parts and will need to be brought into balance with some pruning at planting (see pages 16 and 17, 'The root to shoot ratio'). Commercial growers recommend cutting back the stem to within five or six buds of the graft to balance the above- and below-ground parts. In warm regions where summer temperatures are frequently above 30°C, such severity may be detrimental. Retaining leafy side growths in the trunk will protect the young bark from sunburn, so consider your climate and prune accordingly. A plant with such little anchorage means staking is essential (see page 24, 'Staking'). The tree may take some time to establish; however, once it has settled, prune to encourage wide angled scaffold branches that are spaced at least 60 cm apart. Side growths from these can be thinned to 20 cm apart. Once your framework has been established, future pruning will depend on what cultivar of walnut you have planted.

EVERGREEN FRUIT TREES

Avocado *Persea americana*

Avocados grow and fruit in a surprisingly wide climatic range, from the subtropics to temperate climates. They are frost-sensitive especially when young and are also prone to sunburn on hot summer days. They have brittle wood, but their naturally wide-angled branch attachments are strong enough to carry a heavy crop, but they are not for climbing in. Avocados need little formative pruning but may need excessively vigorous vertical or side growths pinched back to encourage branching. Crossing or crowded growth should also be cleared to keep the canopy open to light and air.

Figure 7.1 Avocados produce millions of flowers but only a few will set fruit.

Always buy a named variety from a reputable grower. Avocados are very susceptible to root rots so buy a tree that is certified free of virus and *Phytophthora* fungus and make sure your planting site is well-drained. They are easy to grow from seed but seedling avocados will not produce fruit for many years, and if they do it may be of inferior quality. Like citrus trees, avocados have a shallow root system so clear all weeds before you plant and mulch well.

Avocados are usually grown as a tree; however, by encouraging side growths and pinching out the apical buds of branches they can be grown as a large shrub. Of course, the amount of fruit produced will also be limited in proportion to its size.

Fruiting wood

Avocados flower at the ends of their freely produced side growths; a simple matter compared to the intricacies of pollination. They bear both male and female flowers on the one tree, but depending on your climate they may not pollinate themselves. Some avocados have flowers that start the day as receptive females, but by the afternoon these have changed to pollen-bearing males. In warm sub-tropical areas there is no overlap of male and female flowers so that an alternate cultivar with the reverse pattern of sex changing flowers is needed for pollination. These are referred to as 'A' and 'B' type avocados and both types must be planted for a crop in such areas. Those growing in cooler climates can be self-fertile due to the fluctuating temperatures that promote the synchronisation of male and female flowers on the same tree at the same time. No matter what region they are growing in, only a few of the millions of avocado flowers will set fruit.

Pruning time

Prune after harvest to shorten long lanky growth. Major renovation of the tree should also be carried out at this time. Dead or diseased wood should be dealt with as soon as you see it.

Maintenance

Once the avocado is established with strong branches, it rarely needs pruning; however, as they are large trees to 10 m and their wood is brittle it is easier to gather the harvest from a 3–4 m tree. Avocados can be cut back extremely hard, even skeletonised, as they will produce strong buds from their old wood, but fruiting will be unlikely in the year following. The major problem with this strategy is that the bark will be unshaded and vulnerable to sunburn. Paint the exposed stems with white water-based paint. A less drastic solution is to remove a branch every few years to keep the tree small and the fruiting wood within reach.

Harvesting

Avocados ripen off the tree and can be stored on the tree for two months. Harvest them later rather than earlier as immature fruits

will shrivel rather than ripen. The fruit stems will start to turn yellowish and wrinkle when they are ready. Always cut your avocado from the tree rather then pull them off. Once the stem attachment is removed from the fruit, rot sets in quickly.

Carob *Ceratonia siliqua*

Carobs have been cultivated for millennia, providing highly nutritious and high-energy food for humans and animals alike. They can be grown as a screening tree to 10 m or encouraged to have a lower branching habit that makes harvesting the pods manually more convenient. Commercial orchards always have trees with a trunk at least a metre high to facilitate mechanical harvesting. Grow them with a central trunk pyramid style (see page 128).

Carobs carry male and female flowers on different trees although there are some hermaphrodite trees that will of course be self-fertile. If you grow your trees from seed

Figure 7.2 Carob. (Photo courtesy of The Food Forest, www.foodforest.com.au)

there will of course be no guarantee what sex they will be. Planting named cultivars will save disappointment.

Although carobs are very drought-tolerant once established, they need reasonable rainfall/irrigation to crop well.

Fruiting wood

Carobs fruit on mature wood about 30 cm in from the edge of the canopy and have the unusual habit of producing flowers directly out of a branch.

Pruning time

Prune after harvest, if at all.

Maintenance

Carobs need little if any pruning. Very crowded and crossed growth that inhibits the penetration of light into the canopy can be thinned out, and branches can be shortened to make harvesting easier. Every few years they will benefit from the removal of a branch to allow the light in. Carobs are very prone to sunburn so after a clean-up prune every few years, paint the stems with a white water-based paint on the most sun-exposed side; that is in the southern hemisphere on the north-west side, in the northern hemisphere on the south-west side. Failure to protect the stems can result in the death of the plant. Although carobs make excellent hedges, pod production will cease with such little light in the interior of the tree.

Harvest

Some carob trees will drop their pods when ripe; others will have to picked or knocked off. Always wait until the green stem that is attached to the pod turns black/brown before picking.

Loquat *Eriobotrya japonica*

Loquats make wonderful tough shade trees and if well-treated, delicious fruit. Seedling trees are fine if all you are after is shade but only named cultivars should be grown if you want to partake of this first fruit of the season. They flower in late autumn to winter, and their generic name means 'wool cluster' in Greek, a very good description of the flowers.

Shape your tree to an open vase for maximum light penetration. If you want both shade and fruit, prune at planting for three to five well-spaced limbs and disbud the trunk to the required height (see page 43, 'Disbudding').

Loquats are self-fertile so only one tree is necessary for fruit. Those grafted to loquat

Figure 7.3 Loquat's generic name *Eriobotrya* comes from the Greek *'erion'* meaning 'wool' and *'botrys'* meaning 'cluster'.

seedling stock can reach to 10 m, while the use of quince rootstock produces a smaller tree.

Fruiting wood

Loquats produce dense clusters of flowers on the current season's growth generally towards the end of the shoot. Although not all the blossoms will be pollinated fruit thinning is essential for good-sized fruit.

Pruning time

After harvest and in late spring to early summer is the best time to prune. Twiggy shaded side shoots can be removed, saving on fruit thinning next year. Any correction of the desired shape can also be performed. Pruning in summer also has the advantage of dwarfing the tree.

Maintenance

Once a framework has been established very little pruning is required. Thinning the fruit clusters in late winter to spring will prevent the tree bearing biennially and improve the quality of the fruit. Loquats do need to be fed well if they are to fruit well so don't neglect the manure or the blood and bone.

Shortening long shoots and removing dead or crowded growth is all that a loquat needs.

Harvest

When the fruit has reached its deepest colour (experience is the only guide) and gives off a pleasant loquat perfume, cut your fruit from the tree. They are easily bruised so handle them gently.

Macadamia *Macadamia integrifolia, M. tetraphylla*

It is so satisfying to know that the most delicious of nuts is also the one requiring the lowest maintenance. They have a huge growing range from the sub-tropics to the warm temperate regions tolerating frosts to −4°C. The *M. integrifolia* hybrids are more effective in warmer areas whereas *M. tetraphylla* copes well with cooler conditions. All, however, need a good supply of water; more than a lemon tree. Most cultivars available are selections or intraspecific hybrids of these species.

Train your macadamia to a single trunk and encourage wide branch angles by removing any side growths growing too close to the main stem (see Figure 4.3, page 50). Keep the main branches about 50 cm apart and cut back the central trunk if it needs some stimulation to branch. Macadamias produce nuts on old and new wood.

Apart from this, leave your macadamia to look after itself!

Figure 7.4 Macadamias.

Olive *Olea europa*

Olives are uncomplaining, very drought-tolerant screening trees that can be utilised in almost any landscape. They can be clipped to a hedge or topiary shape or left to develop into large trees. But if you are after olives to pickle or press, some pruning will be required.

Olives fruit best when their crown is open to air and light so an open vase well-spaced bush/standard or fan espalier are very effective (see pages 126, 132 and 133). Select three to four limbs that will form the scaffold branches and prune off side growths to the height of your proposed trunk. This can be as little as 60 cm from the ground or taller if you want an open standard shape. The latter shape will require a balance between aesthetics and fruit production. The denser the 'ball' on top of the long trunk, the less fruit there will be.

Olives are partially self-fertile with the production of more hermaphrodite/perfect flowers influenced by the trees access to nutrients and water. The dominance of unfruitful male flowers is an indication that your olive will need a little more care. Whichever way you manage your tree the presence of another will improve fruit set.

Fruiting wood

Olives fruit from the leaf axils formed on last season's growth so regular pruning will ensure a plentiful supply of such wood.

Pruning time

Pruning is best carried out just after harvest so not too much vegetative growth is stimulated. Renovating or rejuvenating an old tree can be done in winter to promote the rapid regrowth of vegetative shoots to reclothe the branches.

Maintenance

Olives produce new strong growth from dormant buds on old branches. This characteristic can be useful for renovation work, but the regrowth should be thinned to prevent the canopy becoming dark and dense, rub off unwanted growths as soon as you see them (see Figure 3.4, page 31). They are also prone to suckering at the base of the tree, often as a result of mower damage (see page 46).

Keep the tree open to light and prune for shape and ease of harvest by shortening back longer branches and clearing weak side growths.

Renovation

Olive trees that have become unproductive can be easily, if rather dramatically, renovated. A choice has to be made between its beauty as a large evergreen tree and fruitfulness (see Figure 7.5).

If the choice is for fruit, be brave and have faith. Olives regenerate remarkably well from dormant buds on old wood.

The interior of the tree will be dark and the only fruit production will be at the periphery of the branches (see Figure 7.6).

Start by removing any branches from the center of the tree (see Figure 7.7). Select three or four major scaffold branches and clear out the rest.

Figure 7.5 This mature olive tree, although quietly beautiful, is not fruitful.

Figure 7.7 Remove large branches from the centre of the tree and then the crossed and crowded growth that remains.

Try to prune to a point of growth that is facing away from the centre of the tree (see Figure 7.8).

Figure 7.8 Shorten back the remaining branches so that they will regenerate at the desired height. Prune to a point of growth that is facing away from the centre of the tree.

Figure 7.6 The interior of the tree is dark and crowded with crossed and vertical growth that bears no fruit.

Figure 7.9 Olives can be cut back to stumps and still regenerate as is necessary in this commercial orchard.

Figure 7.10 Ten months later it has completely recovered. The dense new growth needs to be thinned and the suckers removed.

Olives have remarkable regenerative powers (see Figures 7.9 and 7.10).

Such drastic pruning will eliminate that season's crop but can bring the canopy down to a manageable height. After such severe pruning watch out for suckers from the roots and be ready to rub off unwanted growth from the branches. The tree is also now exposed to sunburn so a coat of white water-based paint will prevent the bark from scorching.

Fruit thinning

Olives can be biennial bearing, which is a good crop one year and practically nothing the next. This can be alleviated by thinning the fruit. Allow about five fruit per 30 cm of twig for the largest fruit with maximum oil content. Keep the tree watered in very dry times to ensure that the tree does not drop what fruit it has.

Harvest

The method of harvesting will vary depending on the end use of the olives. Fruit to be pressed for oil can be shaken or knocked from the tree. Pickling olives should be hand-picked to prevent bruising of the fruit. Green olives should be picked as the fruit turns from a dark green to a lighter green. Black olives should be harvested when they are dark but still firm.

White sapote *Casimiroa edulis*

This delicious creamy fruit is unaccountably rare and yet is no harder to grow than an avocado or citrus fruit. Invest in a named

Figure 7.11 White sapote tastes like a smooth rich custard and is as easy to grow as citrus. (Photo Norwood Industries P/L)

cultivar as seedling trees can grow to 20 m and their fruit quality is variable. Grafted trees are a more manageable 5 to 10 m depending on the climate they are grown in. The cooler the climate the smaller they are.

Some cultivars of white sapote can be self-fertile; however, some require a pollinator. Your nursery supplier will be able to give guidance on suitable cultivars.

They are upright trees that need to be pruned to a short trunk and the branches encouraged to wide branch angles with weights or spreaders (see Figure 6.14, page 151). It also tends to grow long unbranched stems so pinch out/cut back branches once they are a metre or so long to encourage side growth. White sapote is well suited to training as a pyramid shaped tree (see page 128).

Fruiting wood

The white sapote bears fruit on the end of new growth arising from a bud produced the previous season.

Pruning time

Prune after harvest to remove growth that has already fruited. Leave the shoots of that season to carry next year's crop.

Maintenance

Once wide-angled scaffold branches have been established, the white sapote needs little regular pruning. Prune for shape and to keep the canopy open to air and sunshine.

Harvest

Some white sapotes change colour to yellow or orange as they ripen; others maintain the fresh apple green of juvenile fruit. Those that do not change colour will feel slightly soft. They can be harvested when not quite ripe as they will ripen off the tree.

8
CITRUS

Citrus fruit *Citrus* spp., *Fortunella* spp.

These warm climate fruit are as ornamental as they are fruitful. Every garden that can grow them should have at least one member of this group, and with some cultivars tolerating down to −5°C, there should be a place for one, even if it is in a pot.

Meyer lemons and kumquats are the most cold-hardy with Lisbon and Eureka lemons being the most tender. Oranges grow surprisingly well in cooler climates but there

Figure 8.1 The gorgeous variety of citrus fruit.

is rarely enough heat in these climes to build the sugars in the fruit. Sour oranges such as the Seville or Bergamot may be alternatives and they are also more cold-tolerant than the sweet oranges.

Always buy virus-tested plants as once these trees are infected, death of the plant and the contamination of the soil will ensue. The tribe are self-fertile so only one plant is required to produce fruit.

Citrus require very little pruning. The exceptions are lemons that need their long lanky growths tamed (see below). Train your tree to three or four major branches once it has become established, shear them as a hedge or train as a fan espalier. Keep their branches well clear of the ground to prevent soil-borne fungi splashing the plant and clear any crossing growth.

Rootstocks

Citrus should always be purchased as grafted plants. Seedlings will not come true to type; that is, the fruit will most likely be inferior and it may take many years to fruit. A grafted tree can be bought in fruit at a year or so old. Always remove the fruit at planting and until the plant is three to four years old.

There are various rootstocks available that confer disease-resistance, tolerance to differing soils and determine the ultimate size of the tree. Citronelle or rough lemon makes a large vigorous tree needing sharp drainage in light soils. The citrange stocks are smaller with better disease-resistance, and the trifoliate stocks are smaller again and are tolerant of disease and heavy soils. The most dwarfing rootstock is Flying Dragon and perfect for citrus in pots. It is slow growing and therefore more expensive, but ideal to dwarf your selected cultivar. A reputable nursery should be able to select what is best for your area and needs, and also ensure that the plant is virus tested.

Fruiting wood

Flowers and fruit are produced from the leaf axils of young growth. Depending on your climate, lemons can flower and fruit year round while most other citrus have one flowering and fruiting period. Kumquats flower in summer while the rest bloom in spring. The time from blossom to harvested fruit will also vary with climate.

Pruning time

In warm, frost-free climates prune any time that takes your fancy. Those who experience frost should not prune until the danger is over some time in late spring. Prune just before the spring flush of growth if the climate is somewhere in the middle.

As most citrus do not need much trimming, prune as you harvest by cutting the fruit from the tree with a length of stem attached. Prune back to the next leaf or more if the growth is becoming straggly.

Fruit thinning

Mandarins are the only one of this tribe to need serious fruit thinning. They can be prone to biennial bearing so thin about every young fruit in four to ensure regular crops.

Maintenance

Always remove crossed and crowded growth to keep the centre of the tree open. Diseased and dead wood should also be dealt with as you see it (see Figure 8.2).

Lemons are the only ones that need regular attention to prevent the branches from becoming too long and leggy. Shorten back the long stems by about a third to an outward-facing bud and thin the old fruiting wood as required (see Figure 5.25, page 139). In hot climates or after severe pruning that exposes once shaded bark to the sun, the major stems should be painted with white water-based paint to prevent sunburn.

Figure 8.2 Citrus gall wasp has caused the swelling on this branch. Cut it out as soon as you see such a growth to prevent it spreading. Destroy the stem as it may still harbour the wasp or its eggs.

Figure 8.3 Dormant buds have been stimulated by hard pruning. Rub off unwanted growth.

Figure 8.4 The regrowth on this hard-pruned kumquat needs little thinning. The yellow leaves are the result of cool winter temperatures that slow the movement of nitrogen in the plant. Warmer temperatures and blood and bone in spring will cure this problem.

Renovation

Citrus have masses of dormant buds beneath their bark so they can be pruned back hard and rejuvenate (see Figure 8.3).

Simply rub off unwanted regrowth that is heading towards the centre of the plant and select one or two to become the new major branches (see Figure 8.3).

The bushy kumquat in Figure 8.4 does not need too much thinning as it is trained as a standard, and dense foliage rather than fruit production is the aim.

Harvest

Limes are the only fruit that drop when they are ripe and not before; however, the rest can be harvested when they have gained almost full colour. The fruit will last a week or so on the tree without loss of quality. The late-fruiting mandarins and tangelos can almost be stored on the tree for a month or more, while the early fruiting mandarins need to be picked as soon as they are ready.

Always cut your fruit from the tree so that the 'button' attachment that holds the fruit to the stem is not damaged. Any break in the fruit's surface will provide an invitation to rot, keep the fruit unbroken and it can be stored for weeks.

Table 8.1 Citrus fruit

Bergamot orange *Citrus bergamia*
Chinotto orange *Citrus aurantium* var. *myrtifolia*
Citron *Citrus medica*
Grapefruit *Citrus x paradisi*
Kaffir or Macrut lime *Citrus hystix*
Kumquats *Fortunella japonica* and *F.margarita*
Lemons *Citrus limon*
Lime *Citrus latifolia*
Mandarine *Citrus reticulata*
Seville orange *Citrus auranticum*
Sweet oranges *Citrus sinensis*
Tangelo *Citrus reticulata x paradisi*

Pineapple guava flower

9
FRUITING SHRUBS

Pineapple guava *Feijoa sellowiana* syn. *Acca sellowiana*

At its best, this is one of the most delicious of fruits carried on a shrub that is almost indestructible. It grows into a perfect evergreen screening shrub or can be easily hedged, although the best fruit production is from an open multi-stemmed shrub open to air and light.

Figure 9.1 The pineapple guava or *Feijoa sellowiana* is easy to grow and delicious. (Photo courtesy of The Food Forest, www.foodforest.com.au)

Seedling plants are often met with and can be satisfactory, if not the best possible. Choose grafted or cutting-grown plants of named cultivars for the best results. Most of these are self-fertile but their yield improves with cross-pollination from another cultivar. Consult your nursery supplier and if they don't know, find one that does, as named cultivars are available from mail-order companies.

Train them to a single stem with five or six major branches, keeping the branches about 1 m above the ground. This makes mulching and feeding their shallow root system much easier.

The delicious flowers that can be used as exotic garnishes are produced on the base of the previous season's wood. Thin and shorten the new growth at harvest to encourage this growth. Apart from the establishment of a framework, very little pruning is necessary except perhaps to tidy long, lanky stems and to improve its shape. Ensure the centre of the shrub does not become too congested by clearing crossed wood once in a while.

The grey-green zeppelin-shaped fruits do not colour when they are ripe which means the birds rarely bother them. When the first fruit starts to fall pick the fruit. It will ripen off the bush in a few days.

Cherry guava *Psidium littorale* var. *longipes*

Cherry guavas are ideal for the domestic fruit garden as they thrive and produce even when neglected. They have a long harvest season and prefer dryish soil conditions rather than wet. Cherry guavas can be satisfactorily grown from seed or cuttings and there are no named cultivars available as yet to my knowledge. They are also a good screening evergreen small tree or shrub when trained to a central trunk and are happy in a large pot. They are self-fertile so only one plant is needed to produce fruit.

The flowers and fruit are produced on the current season's wood so shorten back the branches after harvest by about a third. A tidy-up of crowded growth every now and then is all that is needed to let light and air in to the tree so that it fruits all over, not just at the tips. Pick the fruit when it is fully coloured and slightly soft once or twice a week for the best fruit.

Tamarillo, tree tomato *Cyphomandra betacea*

This enormously productive, short-lived evergreen shrub supplies fruit through the late autumn and winter when fruit is scarce. It is self-fertile, quick growing and will yield a harvest even in its second year. Plant them in a sheltered position as the wood is weak and easily damaged.

Tamarillos can be grown reasonably reliably from seed although cutting-grown plants are bushier and therefore more fruitful. Encourage seedling-grown plants to branch by removing the apical bud at about 1 m from the ground to promote more side growth.

The fruit is carried at the ends of last year's wood so trim it lightly after harvest to keep the growth close to the main stem. As they are so short-lived, take cuttings of your plant about every three years to act as replacements for the parent tree in five to six years time.

Figure 9.2 The cherry guava fruits even under adverse conditions. (Photo courtesy of www.greenharvest.com. au)

Figure 9.3 The tamarillo or tree tomato provides fruit when little else is available.

Pepino *Solanum muricatum*

Depending on your climate, this small sub-shrub can be grown as an annual in cold areas or as a short-lived perennial, at least two to three years, in frost-free climates. Perfect in a pot in a sheltered position, it may need some support to keep it upright in more open situations. Commercial growers train them to a short 1 m trellis. They can be grown much like a tomato. Unlike tomatoes they do require some pinching out of growing shoots (see page 32) to increase the side growths and therefore the yield. A balance must be reached between the density of fruiting stems and the leaves' access to light.

The juicy semi-sweet fruit with the texture of a melon are borne in clusters. Always pick the largest of the fully coloured fruit first as the smaller ones will continue to grow in size.

Figure 9.4 The short-lived pepino provides juicy semi-sweet fruit. (Photo Norwood Industries Pty Ltd)

10
BERRY FRUIT

Berries are always best when picked fresh off the plant. They are delicate creatures that are easily damaged. The most popular berries are described below; however, there are others that should be mentioned. Cranberries (*Vaccinium macrocarpon*) require a large marshy area to make them worthwhile to grow. They are a sprawling groundcover that bears its fruit on upright side growths. The twining stems and the fruiting growth need thinning to prevent them from becoming overcrowded. Small-scale landcapes are unlikely to produce more than a few grams per square metre even under ideal conditions; far too little for a useful harvest. There is the added complication that cranberries are usually 'float harvested'. The field is flooded and the ripe berries float on top of the water where they are raked off. This is not a viable option for any except the fanatic.

The carissa (*Carissa grandiflora*) is a variable-tasting berry that is produced on new growth from a thorny, slow-growing, pretty evergreen shrub with perfumed white flowers. Often planted as a hedge, it is extremely attractive; however, there are not (to my knowledge) any named cultivars so the fruit may range from the tasteless and mealy, to juicy and sweet. If

you want to take the punt and plant it, prune it back after harvest

The ugni (*Ugni mollinae*) is an intensely flavoured berry making aromatic jams and jellies or a piquant snack. It functions well as a hedge that requires some shade in warm climates and moisture in dry times. Flowering on new growth, the berries should be picked when they have lightened from burgundy brown to dull dark red. Ugni jam and jelly was a favourite of both Queen Victoria and Edna Walling – recommendation enough. Trim it back after harvest for a handsome glossy evergreen hedge that looks like a darker version of English box.

The Gogi berry (*Lycium barbarum*) is known under many different common names such a wolfberry, Duke of Argyll's tea-tree or Chinese boxthorn. Always purchase a plant labelled with the botanic name to avoid confusion. This deciduous twining, suckering shrub grows so easily in sun or semi-shade in all but waterlogged soils that it is listed as a noxious weed in some areas. Check with your local authorities before planting.

The berries are famed for their health effects, while the leaves, shoots and the bark from the roots are also used medicinally. Generally prune out some older wood and shorten back the growth produced the previous spring just to keep it tidy. This growth will produce the next crop so don't be too severe. In frost-free areas, prune after harvest or for frosty climes in spring. Shake or gently pluck the berries when they are fire-engine red and slightly soft. Once established, the Gogi can be pruned into old wood and it will resprout. They respond well to coppicing (see page 84) every six to seven years to rejuvenate the bush.

Blueberry *Vaccinium* spp.

Blueberries are ornamental, delicious and easy to manage; however, they are very fussy when it comes to soil. Whether you grow the deciduous highbush berries or the semi-deciduous to evergreen tetraploid hybrids, blueberries demand acid soil, the hybrids being more tolerant of alkaline conditions. Perfect drainage and constant soil moisture are other requirements and if these conditions can be fulfilled, they are the perfect low-maintenance berry shrub. Mulching with acidifying materials such as oak leaves, pine needles or rhubarb leaves will protect the surface root system.

Blueberries are self-fertile but both the crops and the berries will be larger when more than one cultivar is grown. Named cultivars are grown from cuttings so stems can be cut to the ground and the new suckering growth will still be of the desired cultivar.

Plant your blueberries and let them establish; they need no pruning at planting, but the flowers should be removed for the first two years while the plant settles in.

Figure 10.1 Blueberries flower and fruit from growth formed the previous season.

Fruiting wood

Blueberries flower and fruit from growth that was produced the previous year (see Figure 10.1).

A strong new cane will fruit along its stem in its second season, then bloom on side growths for the next two years. After this time the cane will be spent and should be pruned out.

Pruning time

After harvest in late autumn is a good time to assess the state of your bushes and prune unwanted growth. Prune off the ends of fruiting canes in spring to effectively thin the harvest, so that larger berries are produced. Masses of tiny berries make a disappointing crop.

Maintenance

The spent four-year-old stems should be removed either down to the ground or to a strong side growth to encourage new canes. Old twiggy, droopy, crossed or crowded

Figure 10.2 Four-year-old stems can be pruned out after harvest. Cut them at their base or back to a strong side growth (1). Vigorous new stems will carry a crop next season (2) and the year after fruit on new side growths (3). Very vigorous new growth can be tied down to improve the stem's fruitfulness (4).

growth should be removed to allow light and air into the canopy, leaving stems that are at least the width of a pencil. If there is not much of this growth, cut back the weak stems by about a third and perhaps review the amount of nutrition you are supplying. Blueberries benefit from a good dressing of rotted manure. Strong new vigorous canes can be tied down to encourage more fruitful wood and stop the plant running skywards. Such vertical stems will produce a mass of vegetative growth with little fruit so tie the stem down to its neighbour with soft plant ties (see Figure 10.2).

Harvest

Blueberries do not ripen off the bush so ensure that they are fully ripe before picking. The berry should be completely blue with no green shading near the stem. Gently twist the fattest and darkest berries from their stem to avoid tearing the skin. Torn skin equates to a mouldy berry if it is kept longer than a day. The fruits in each cluster will ripen at differing times so pick every other day.

Currants *Ribes* spp.

Currants are not just currants; they can be red, white or black. Red and white currants are essentially the same except for their colour and are pruned in the same manner. Black currants are treated differently; they have different fruiting wood, prefer cooler winters and more water-retentive soil than the reds and whites, although both require soils rich in organic matter. All currants are grown from cuttings and purchasing named

Figure 10.3 Who needs rubies when you have red currants!

cultivars will ensure the best fruit. Currants need to be netted or grown in a berry cage if you are to pick any crop; birds love currants.

Red and white currants *Ribes sativa, R. rubrum*

These medium-sized bushes to 1.5 m have brittle wood so plant them away from strong

Figure 10.4 Red and white currants fruit from long-lived spurs.

winds. They can be grown as standards, espaliers or as an open vase bush on a short 30 to 40 cm trunk with a base of three to four branches (see page 130). Even if they are a bush, the basic principles are the same. Standard-shaped bushes can be grafted onto a flowering currant that will form the stem, and will need a sturdy stake for support.

Fruiting wood

Red and white currants fruit from two- to three-year-old spurs in much the same way as apples and pears (see Figure 10.4).

These spurs are long lived and so a permanent structure of branches can be developed. A neat bush or espalier shows off the beauty of these jewel-like fruit (see Figure 10.3).

Pruning time

In the early stages of establishment, winter pruning will promote vigorous branches to provide a basic framework. The fruiting wood real or proposed should be pruned in summer. Prune the new side growths to about 15 cm as the fruit is ripening and then shorten them back again to about two buds in winter.

Maintenance

The summer and winter pruning should remove about half the annual growth, some spur thinning may be needed and bush trained currants can benefit from the removal of a branch every four to five years. As with all fruiting plants, keep the centre of the bush open, and always remove any dead, diseased, crossing or crowded growth.

Harvest

Red and white currants should be clipped from the plant with scissors or snips so that the fruiting spurs are not damaged. The bunches of fruit can hold on the bush for a few weeks without dropping the berries, so taste and decide for yourself when they are ready.

Black currants *Ribes nigrum*

Black currants have had bad press about the smell of their fruit and leaves. Suffice it to say, they smell like black currants and provide more vitamin C than oranges, make splendid jelly and highly nutritious cordial. They are not a fruit to be enjoyed straight off the bush. They are grown as a stool or coppice (see page 84, 'Coppicing'), and should be planted deeply so that shoot growth/suckers can develop from beneath the soil. Once planted in winter, prune back the stems to two buds from ground level (see Figure 10.5).

Fruiting wood

The most productive wood is formed the previous season with two-year-old wood still fruiting, but not as vigorously. Wood that is three years old should be cut to the ground or to a strong growing side shoot.

Pruning time

Once the bush is well established at three years old, the fruiting wood can be cut out at harvest and the berries simply stripped off the stems. Winter is the usual pruning time when the one-year-old wood can be thinned, leaving the thickest stems and removing any

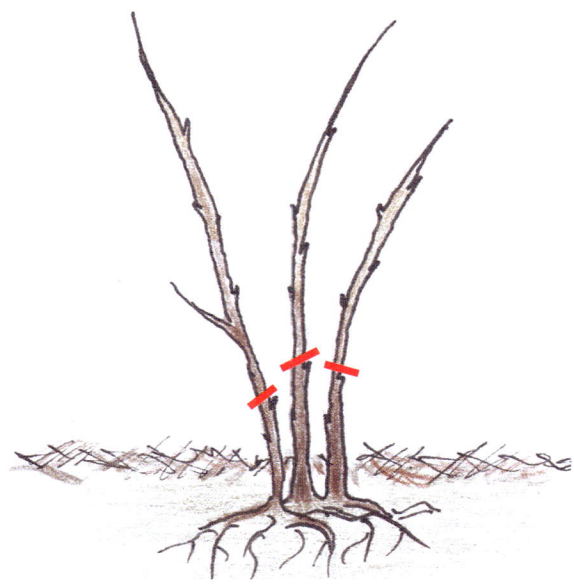

Figure 10.5 Plant black currants deeply to encourage new growth/suckers from the base. In winter cut the stems to two buds above ground level.

twiggy spindly growth. Any stems three years old or more should also be removed; there should not be any of this wood if all the stems carrying fruit have been cut down at harvest.

Maintenance

Black currants need to have their wood renewed constantly so old wood is pruned out to make way for new suckers to form the future fruiting wood. Pruning out fruiting stems and thinning the new growth will keep the bush productive (see Figure 10.6).

Harvest

These currants ripen unevenly so it is best to wait until the berries at the top of the cluster have fallen before harvesting. As they are a culinary berry, it does not matter too much if some of the lower berries are unripe. Unlike

Figure 10.6 Once the plant has established, prune out fruiting wood to ground level or to a strong side growth at harvest. In winter, thin the newest growth selecting the strongest stems that will carry next year's fruit. Cut out the weak.

Figure 10.7 Gooseberries are gorgeous to look at, but NEVER eat an unripe one. (Photo by Norwood Industries Pty Ltd)

the reds and whites, the berries do not hold on the bush, so harvest as soon as you notice the first of the cluster dropping.

Gooseberry *Ribes* spp.

Gooseberries are a favourite cool-climate home-garden fruit. The bushes themselves are very frost-hardy but the fruits can be damaged and are prone to splitting in wet weather. They can be grown as a stool or coppice (see 'Coppicing', page 84), or as bush on a short trunk.

Fruiting wood

Gooseberries fruit mainly on spurs that are two to three years old and, to a lesser extent, on one-year-old side growths. When grown as a coppice, cut out all wood more than three years old and thin the one-year-old growth that will bear spurs in the coming two years. Shorten long and lanky one-year-old wood in

summer to encourage spurs and ensure that it will not drag its fruit load on the ground.

Pruning time

Prune in winter when it is easy to see the dark peeling stems that indicate old wood. Thinning the one-year-old stems allows light and air into the plant and makes picking more comfortable – gooseberries have ferocious thorns!

Harvest

If you want to make gooseberry preserves, pick the berries when they are underripe. If they are for fresh eating, wait until their colour has developed fully – an unripe gooseberry is savagely sour!

Strawberries *Fragaria x ananassa*

Strawberries are a short-lived herbaceous perennial very prone to virus and moulds so always buy certified virus-free stock and replace your plants every four years. Never

Figure 10.8 Surely the most popular and beautiful berry.

Figure 10.9 Strawberries reproduce from runners. Use them to increase your stock, but for the best crops remove the runners so that the plant's energy goes into fruit production.

plant strawberries in the same place twice. Diseases build up in the soil so change their position in the landscape. This is a good argument for growing them in pots where the soil can be replaced completely and the containers can be washed thoroughly. A position catching the morning sun is ideal as this will dry the morning dew quickly so fungal problems can be minimised. Netting to protect the fruit from birds is essential.

There are two main types of cultivated strawberries; the summer flowering berries and the day-neutral ever-bearing strawberries. The summer flowering plants can be renovated replacing older plants with fresh runners every two to three years, while the everbearers are best replaced completely after three to four years.

Fruiting wood

Strawberries produce their flowers, fruits and leaves from their crowns like all herbaceous perennials. The flowers are formed in the crown the preceding summer, so water and fertilise well after harvest to ensure next year's crop.

Pruning times

Pinch out the first flowers produced after winter planting for about two months so that the plants can establish. Cut back all the foliage after harvest.

Maintenance

Summer strawberries can and should be renovated to prolong the life of the strawberry bed. They reproduce by means of runners (see Figure 10.9), and it stands to reason that establishing a new plant runner and good berry production are mutually exclusive activities. Unless you wish to increase your stock of plants, remove all runners so that the plants energy can be directed to berry production.

Straight after harvest cut back all the leaves and any remnant flowering stems with hedging shears, or just grab the leaves and chop them off with the secateurs. The plants should look something like an echidna. Now is the time to fertilise and water well so that new leaves will be produced to feed the crown that develops the embryos of next year's flowers.

Harvest

Harvest your berries when they are completely coloured and glossy.

11

CANE BERRIES

Raspberries *Rubus idaeus, R. idaeus* var. *strigosus*

Raspberries fall into two groups: the summer-fruiting and the autumn-fruiting that can also produce in summer; but more on this later. The summer-fruiting raspberries are generally of better flavour, and by selecting different varieties can provide a harvest from early to just after mid summer.

All raspberries are best trained to a trellis or some sort of support so that the canes can be securely tied to prevent damage and don't become a tangled mess! A trellis with a wire at 70 cm high and then another strained at 1.6 metres high will suffice. Plant your berries in winter slightly deeper than they were in the pot and cut them back to 20 cm. In their first season they should not be allowed to fruit. The plant's energy is best diverted into the production of a strong root system.

Fruiting wood

Summer-fruiting raspberries fruit on wood that was produced the previous season. They are called primocanes in their first year and floricanes as they enter their second, fruiting season. Once a cane has borne fruit it should be removed (see Figure 11.1).

Autumn-fruiting raspberries fruit in late summer on growth that was produced that season and then again the next summer on the same canes if they haven't been pruned out.

Figure 11.1 It is easy to tell the difference between the new and the older wood. The spent floricanes on the right are dark. The future fruiting canes are paler on the left. Prune out the new growth that is weak.

Pruning time

Pruning can be carried out between the end of harvest and before growth starts the next spring. Summer-fruiting raspberries can have the spent canes that have fruited (old floricanes) removed straight after harvest; however, in warm climates the old floricanes can provide some shade for the new growth, next year's fruiting wood or primocanes.

Extra leaf coverage from the old floricanes will also feed the developing primocanes.

Most pruning is carried out in autumn to winter, with the thinning of weak or crowded primocanes as they develop over the season (see Figure 11.2).

Autumn-fruiting berries can be cut to the ground completely after harvest every second

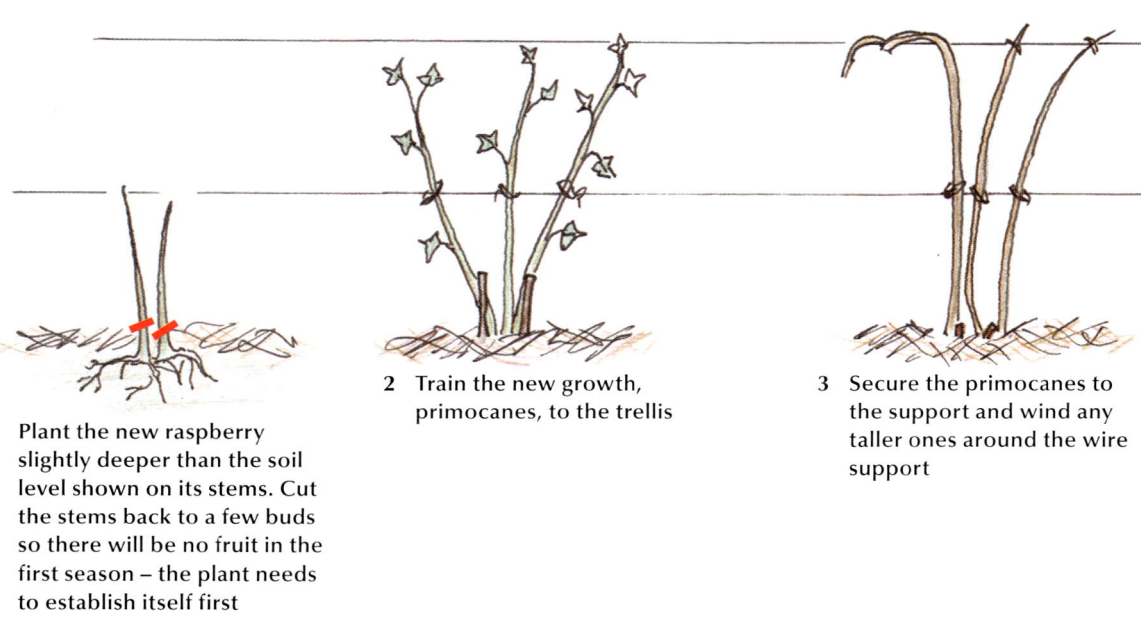

1 Plant the new raspberry slightly deeper than the soil level shown on its stems. Cut the stems back to a few buds so there will be no fruit in the first season – the plant needs to establish itself first

2 Train the new growth, primocanes, to the trellis

3 Secure the primocanes to the support and wind any taller ones around the wire support

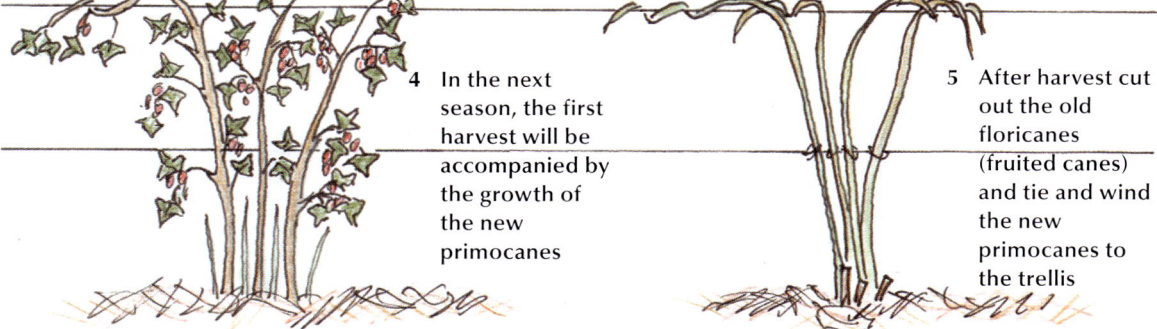

4 In the next season, the first harvest will be accompanied by the growth of the new primocanes

5 After harvest cut out the old floricanes (fruited canes) and tie and wind the new primocanes to the trellis

Figure 11.2 Training and pruning summer-fruiting raspberries.

season to produce two crops in the one season. Many find it easier, however, to enjoy the autumn crop and cut down the entire plant after harvest every year. Growing both summer- and autumn-fruiting raspberries can yield fruit for up to six months.

Maintenance

How you maintain the fruiting wood on raspberries depends on whether you are growing summer-fruiting or autumn-fruiting cultivars. Summer raspberries have one generous crop per year. Autumn raspberries can be pruned to produce two crops per year. However, most growers cut down their autumn fruiting raspberries completely straight after harvest. It is up to the pruner's preference.

This process is best explained visually (see Figure 11.2 for summer-fruiting raspberries and Figure 11.3 for the autumn cultivars).

There are some general simple rules to follow. Remove any cane that has fruited from its tip and its stem. Thin the new growth, primocanes, selecting the strongest and

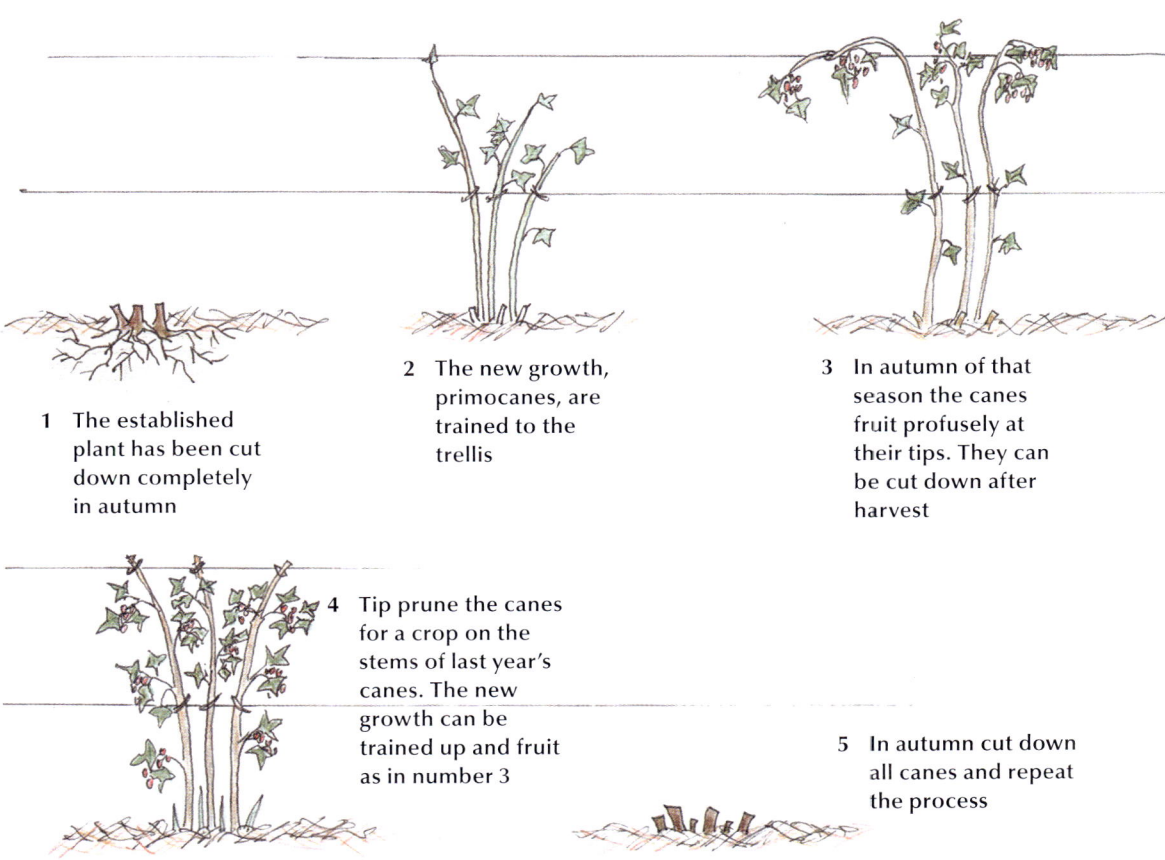

1 The established plant has been cut down completely in autumn

2 The new growth, primocanes, are trained to the trellis

3 In autumn of that season the canes fruit profusely at their tips. They can be cut down after harvest

4 Tip prune the canes for a crop on the stems of last year's canes. The new growth can be trained up and fruit as in number 3

5 In autumn cut down all canes and repeat the process

Figure 11.3 Autumn-fruiting raspberries.

cutting out any weak or damaged canes through the growing season leaving 15 cm between canes.

As the primocanes grow, tie them to the trellis and twist them around the top wire if they grow taller. Primocanes can have their tips removed to stop them flapping about in the wind, but the longer the cane, the more fruit will be produced and they are easily secured by winding them around the trellis's wire.

As always, remove dead, damaged or diseased wood as soon as you see it.

Harvest

Raspberries ripen over a period of four to five weeks for each cultivar so pick your berries daily. Gently squeeze the berry and pull to detach it from its stem.

Bramble berries *Rubus* spp. and hybrids

Bramble berries include the more civilised members of the blackberry clan. They are easy to manage especially the thornless cultivars. The cultivated and semi-tame blackberries and their hybrids most available are loganberry, boysenberry, Marion berry, silvanberry and youngberry, and demand little from the grower in comparison to the harvest.

Bramble berries are naturally trailing plants that take root wherever their shoots graze the soil, so keep them off the ground; a neglected brambleberry is a very scary sight! They should be trained to a trellis with wires spaced roughly 50 cm apart to a height of 2 m and planted about 2 m apart.

Figure 11.4 The cultivated and semi-tame blackberries and their hybrids are best trained well above the ground. Bramble berries can be trained to a trellis by bundling a few canes together and winding them around the wires.

Plant out the berries in winter and cut them back to two to three buds to prevent fruiting and to get them established the first year. Once the new canes grow from the base there are several ways to attach the canes to the trellis. They can be fan trained with 25 cm between the canes, bundled two to three canes together and wound around the wire (Figure 11.4), or woven through the wires (see Figure 11.5).

If you wish to fill a long space at a lower height, train them to a 1 m to 1.2 m high trellis; over a distance of 4 m, they can make an effective low screen (see Figure 11.6).

Fruiting wood

All bramble berries fruit on canes that were grown the previous season. They are called primocanes in their first year and floricanes as they enter their second, fruiting season (see Figure 11.7). Once a cane has borne fruit, it should be removed.

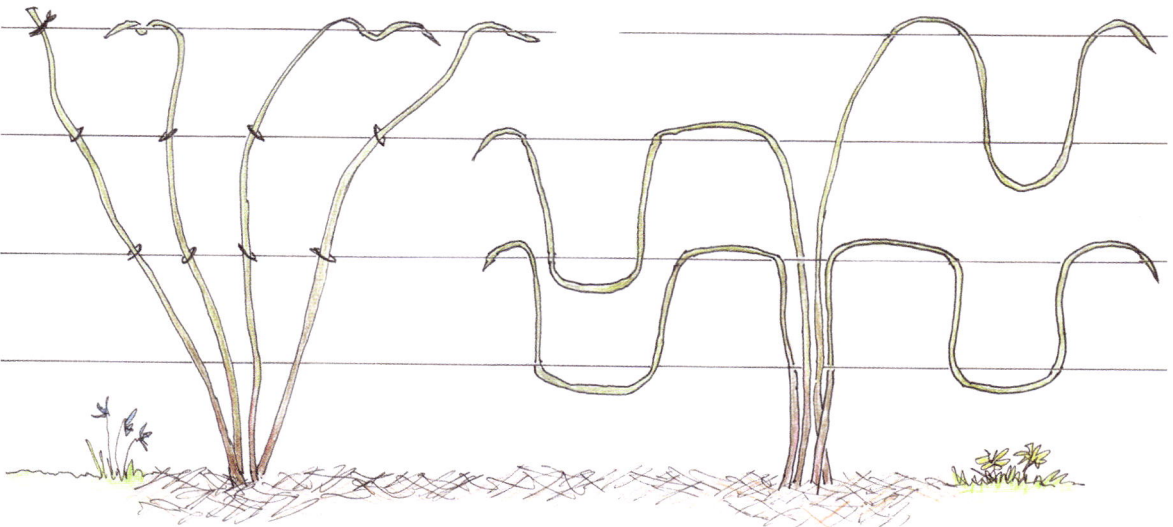

Figure 11.5 Bramble berries can also be fan trained or woven through the wires.

Pruning time

The spent floricanes can be removed directly after harvest and the new primocanes trained to the trellis. It does not matter if this job is delayed as the primocanes can sprawl on the ground until there is time to weave/bundle them on to the support.

Once established, the new growth or primocanes can be cut back lightly in late

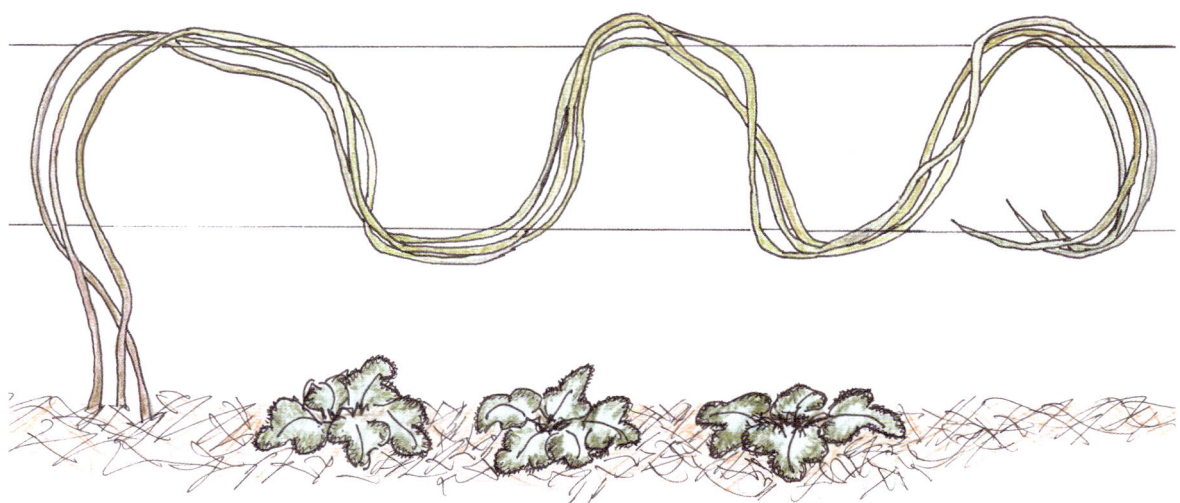

Figure 11.6 Bramble berries can be trained to a 1 m high trellis over a 3–4 m distance and make an effective screen. Select three of the strongest primocanes and bundle them along the wires; cut out the rest.

Figure 11.7 The old floricanes are dark (right) and primocanes, next year's fruiting stems, are green. Cut out the old canes completely.

Figure 11.8 Bramble berries ripen unevenly so pick the fruit every day. The darker the berry the riper it is.

spring so they don't get tangled with the floricanes. It will also encourage regrowth with meristems/nodes (see page 2) packed more closely together. This will keep the plant more compact and with more buds available for the production of flowers and fruit. This is effective in controlling vigorous cultivars on very fertile soils. Assess your plant's strength (or weakness) and act accordingly.

Maintenance

Pruning out spent floricanes and tying in primocanes is the major pruning and

training requirement. As the plant becomes more established, select the strongest six to eight primocanes per plant and prune out the rest. In very vigorous cultivars any side growths produced by the primocanes can be shortened back to two to three buds.

Keep your bramble berry well-watered for the juiciest harvest.

Harvest

Pick the fully coloured fruit daily, gently squeezing and pulling it from the canes. Bramble berries ripen unevenly so can be harvested for several weeks (see Figure 11.8).

GLOSSARY

Acid soils Soil with a pH value less than pH6.5. Calcium and magnesium will be lacking.

Adventitious roots Roots that arise from a piece of plant rather than from a seed, e.g. cuttings have adventitious roots.

Aerial roots Roots that develop on plant parts that are above ground, surviving light and air. In some climbers they are mechanisms that attach the plant to a solid surface, e.g. ivy.

Alkaline soils Soils with a pH value greater than pH7. They will be lacking in iron, but rich in calcium and magnesium.

Apical buds Buds that are formed at the very tip of a stem.

Biennial bearing Plants that fruit profusely one year and fail to crop the following year.

Bulbous plants Plants that grow from a fleshy root or stem, generally monocots with strappy leaves.

Canopy The area that is covered by a plants stems and leaves, like the extent of an umbrella.

Central leader The central stem or trunk of a tree.

Cordons A style of espalier where a tree is trained to a single stem. They are often planted at an approximate 30° angle from the vertical.

Cultivar A plant that has been propagated by gardeners or nurserymen to display specific attributes. Persimmon 'Fuyu' and Persimmon 'Dai Dai Maru' are both cultivars of the wild species *Diospyrus kaki*.

Dicot Flowering plants that originate from a seed with two seed leaves, having a cambium layer forming a ring just under the outer stem (see Figure 1.2). They have leaves with branching veins.

Dioecious Plants that bear flowers of a single sex only. A male plant and a female plant are required to produce fruit, e.g. holly and pistachio.

Dormant/dormancy A stage in the plants lifecycle where no growth occurs, e.g. winter for an apple tree.

Grafting Where the cambium layers of two plant stems are cut (naturally or by design) and matched together to grow into one joined stem. This enables the production of grafted trees (e.g. cultivars of fruit trees) and ensures the strength of pleaching techniques.

Intercalary meristem A point of growth that is situated above the root system and gives rise to grassy or strappy leaves.

Lateral branches Branches that grow to the side of a larger stem and promote the thickening of that stem.

Leaf nodes A small swelling or bump on the stem where the leaf is produced. It is this growth point or meristem that gives rise to the leaf and possibly subsequent branches on the plant.

Meristems A growth point or bud that has the potential to produce a leaf, stem, branch, root or flower through actively growing cells. Meristems can be apical (the top of a stem), axillary (at the side of the stem) or intercalary (between the roots and the leaves). These growth points can be stimulated to grow through pruning.

Monocot A flowering plant that originates from a seed with a single seed leaf, and has leaves with parallel veining, e.g. corn, *Knifophia* and grasses.

Monoecious A plant where separate male and female flowers are carried on the one plant, e.g. pumpkins and hazelnuts.

Organic matter Anything that was once alive or the product of something once alive. They may be high in nutrients like manures or blood and bone or high in carbon like newspapers or straw. A combination of such materials at a ratio of 25 to 30 parts of carbon to one part of nitrogen together with water and oxygen is necessary for successful compost.

pH test A simple soil test that determines the pH level of soil, i.e. whether the soil is acid or alkaline. The pH level will determine to a large degree what plants are suitable for your soil.

Rootstocks The roots and lower part of the trunk on grafted plants that can determine the vigour, size and disease resistance of the desired plant (scion wood) that is grafted onto it.

Scaffold branches Branches arising from a trunk that support the fruiting and flowering wood.

Self-fertile A plant that will bear fruit without pollination from another plant.

Soil aggregates Crumbs of soil made up from the mineral and organic soil particles that determine soil structure. Stable soil aggregates are essential for a soil that can store both air and water, that is neither rock-hard nor caked sand. Soil aggregates should remain stable even after heavy rain.

Sports A branch, shoot or flower that is spontaneously different from the parent plant. For example, a pink rose may produce a sport that produces white flowers.

Spurs Long lived woody outgrowths that produce flowers and/or fruit.

Stolons A branch or stem that runs along the ground and takes root at its nodes or meristems, e.g. Kikuyu grass and strawberry runners.

Tap root The primary root characteristic of seedlings that grows straight down into the soil to stabilise the young plant.

Tendrils A modified plant part that coils around other plants or supports in order for the plant to climb.

Vegetative growth Plant growth that is made up of stems and leaves only – no flowers or fruit.

REFERENCES

Ballinger R, Ballinger J and Swan H (1983) *Fruit Gardening in South-Eastern Australia*. Caxton Press, Christchurch, NZ.

Baxter P and Tankard G (1990) *Growing Fruit in Australia*. Macmillan, Melbourne.

Beales P (1985) *Classic Roses*. Harvill Press, London.

Blazey C and Varkulevicius J (2006) *The Australian Fruit and Vegetable Garden*. The Diggers Club, Dromana, Victoria.

Buscher FK and McClure SA (1989) *All About Pruning*. Ortho Books, San Fransisco.

Chatto B (2000) *Beth Chatto's Gravel Garden*. Bloomings Books, Melbourne.

Edmunds A (YEAR) *Espalier Fruit Trees, Their History and Culture*. Horticultural Press, Carlton, Victoria.

Glowinski L (1997) *The Complete Book of Fruit Growing in Australia*. Thomas C Lothian, Melbourne.

Handreck K (1993) *Gardening Down-Under: A Handbook for Enquiring Gardeners*. CSIRO Publishing, Melbourne.

Handreck KA and Black ND (1994) *Growing Media for Ornamental Plants and Turf*. University of New South Wales Press, Randwick.

Harkness J (1978) *Roses*. JM Dent and Sons Ltd, London.

Hitchmough J (1994) *Urban Landscape Management*. Reed International Books, Australia.

Kilpatrick DT (1968) *Pruning for the Australian Gardener*. Rigby Ltd, Melbourne.

Lawson-Hall T and Rothera B (1995) *Hydrangeas: A Gardener's Guide*. BT Batsford Ltd, London.

Lloyd C (1983) *The Adventurous Gardener*. Penguin, London.

Macoboy S (1982) *Trees for Flower and Fragrance*. Lansdowne Press, Sydney.

Macoboy S (1982) *Trees for Fruit and Foliage*. Lansdowne Press, Sydney.

Mawley E (1909) *Handbook on Pruning Roses*. 2nd edn. The National Rose Society, Herts., UK.

Reich L (1999) *The Pruning Book*. The Taunton Press, Newton, CT, USA.

Toothill E (Ed.) (1984) *Dictionary of Botany*. Penguin, London.

INDEX

branch collar 40–2
branch placement 19
branches of varying vigour 127
Brassica oleracea 22, 23
breba crop, figs 154, 155
broad beans 33
broccoli sprouting 33
broom 55, 76
Brugmansia versicolor 54
buddha grass 66
buddleja 54, 55, 95, 99
Buddleja spp. 54, 55, 95, 99
bulbous plants 60
Buxus microphylla 95
Buxus sempervirens 95, 99

cabbage 11
cabbage roses 70
Cabernet Sauvignon 115
Calamagrostis spp. 61
calendula 33
Callistemon spp. 54, 95, 101
Calothamnus quadrifida 54
camellia 54, 95
Camellia spp. 54, 95
candytuft 33
cane berries 193–6
cane pruning 114, 115–16
canopy removal 53–4, 80
capsicum 33
Carex spp. 61
Carissa grandiflora 185
carob 170–1
    hedge 95, 170
carpet roses 69, 72, 74, 75–6
Carpinus betulus 95, 97, 99
Carrissa grandiflora 96
Caryopteris clandonensis 67, 95
Casimiroa edulis 175
Castanea sativa 83, 97, 153
Casuarina torulosa 95, 99
catalpa 83
Catalpa bignonioides 83
catmint 66
centifolia roses 70
Centranthus ruber 65
Ceonothus spp. 95

Ceratonia siliqua 170–1
cerinthe 33
Cestrum spp. 54
Chaenomeles japonica 95
Chardonnay 115
cherry 150–3
    fan training 132
cherry guava 182
cherry pie 96, 101
chestnut 153
    pleaching 97
    pollarding 83
chilli 33
Chinese gooseberry 117–18
Chinese lantern 101
chinotto orange 179
chloroplasts 5, 6, 58
chocolate vine 109
Choisya ternata 54, 87
Chorizandra spp. 64
Chrysanthemum frutescens 101
cineraria 33
Cistus spp. 55, 96
citrus 177–9
    hedge 95, 96
    renovation 140–1
    rootstock 126, 177
    weather damaged 122
Citrus auranticum 179
Citrus bergamia 179
Citrus hystix 179
Citrus latifolia 179
Citrus limon 179
Citrus medica 179
Citrus reticulata 179
Citrus sinensis 179
Citrus spp. 95, 96, 177
Citrus × paradisi 179
Clb Gold bunny rose 79
Clb Iceberg rose 79
clematis 112, 113, 119–21
Clematis alpina and cultivars 113
Clematis aristata 113
Clematis armandii and cultivars 113
Clematis cirrhosa and cultivars 113
Clematis florida 113
Clematis montana and cultivars 113
Clematis tangutica and cultivars 113

Vaccinium spp. 186–7
valerian 65
variegated plants 49, 57
vascular bundles 1, 2
Verbena × *hybrida* 58
Veronica 65
Viburnum odoratissimum 54
Viburnum opulus 24, 54, 55
Viburnum plicatum 54
Viburnum tinus 87
Viburnum × *burkwoodii* 55
Vietnamese mint 33
viola 33
Virginia creeper 110
Vitis spp. 113
Vitis vinifera 114

walnuts 21, 167,168
water logging 9
water retention 13
watercress 33
weed seeds 12, 13
weeping branches 107, 108
weeping roses 68, 78, 103
weeping trees 103

Weigelia florida 55
weights, use of 127, 128, 150, 151
Westringia fruticosa 95, 99
whip 97
white currants 187, 188, 189
white sapote 175–6
Williams (syn. Bartlett) pear 133
willow 46, 83, 95, 101, 103, 107
willow leaf hebe 95
wind 8
winter pruning 26–7
wisteria 97, 113, 114
Wisteria floribunda 114
Wisteria spp. 97, 113, 114
wounds 29–30

xylem 1, 2, 4, 5

yarrow 66
yew 95, 99
Youngberry 196

zeolites 9
zinnia 33

DISCARD
E.C.L.